Analysis of renewable energy integration options in urban energy systems with centralized energy parks

Analysis of renewable energy integration options in urban energy systems with centralized energy parks

Vom Promotionsausschuss der
Technischen Universität Hamburg-Harburg
zur Erlangung des akademischen Grades

Doktor-Ingenieur (Dr.-Ing.)

genehmigte Dissertation

von

Ricardo Peniche Garcia

aus
Merida, Mexiko

2017

Bibliografische Information der Deutschen Nationalbibliothek
Die Deutsche Nationalbibliothek verzeichnet diese Publikation in der Deutschen Nationalbibliografie; detaillierte bibliografische Daten sind im Internet über http://dnb.d-nb.de abrufbar.
1. Aufl. - Göttingen: Cuvillier, 2017
Zugl.: (TU) Hamburg-Harburg, Univ., Diss., 2017

1. Gutachter: Prof. Dr.-Ing. Alfons Kather

2. Gutachter: Prof. Dr.-Ing. Günter Ackermann

Prüfungsausschussvorsitzender: Prof. Dr.-Ing. Gerhard Schmitz

Tag der mündlichen Prüfung: 07. Juli 2017

Umschlaggestaltung: Nayane Nathalie de Souza Hablitzel

Für den Einband wurden Kartendaten aus OpenVectorMaps.com verwendet

Nonnenstieg 8, 37075 Göttingen
Telefon: 0551-54724-0
Telefax: 0551-54724-21
www.cuvillier.de

1. Auflage, 2017
Gedruckt auf umweltfreundlichem, säurefreiem Papier aus nachhaltiger Forstwirtschaft.

ISBN 978-3-7369-9608-3
eISBN 978-3-7369-8608-4

Acknowledgments

The contents of this thesis are based on the research work carried out during my stay as a researcher at the Institute of Energy Systems (IET) of the Hamburg University of Technology. The work was conducted as part of the project TransiEnt.EE - *Transientes Verhalten gekoppelter Energienetze mit hohem Anteil Erneuerbarer Energien*, which was financed by the German Federal Ministry for Economic Affairs and Energy (BMWi).

I would like to thank Prof. Dr.-Ing. Alfons Kather for his trust, for the opportunity of working at the IET, and for his frequent and valuable feedback during the preparation of this thesis. It is truly an honor to have him as *Doktorvater*. Thanks also to Prof. Dr.-Ing. Günter Ackermann and Prof. Dr.-Ing. Gerhard Schmitz for their genuine and active interest in the project and for being part of the evaluation committee of this thesis.

My gratitude and admiration to my project colleagues Lisa Andresen and Pascal Dubucq. It was a pleasure working together with such capable engineers on the project and the development of the TransiEnt model library. For the many meetings, coffee breaks and hours of technical discussions, I thank you both. Many thanks also for their time and project-related advice to the members of the TransiEnt.EE Advisory Board and to Dr. Johannes Brunnemann from XRG Simulation GmbH. Also thanks to all students that did their thesis under my supervision for their trust and their contributions to tackle specific issues related to the project.

Thanks to my colleagues from the IET for their feedback and their remarkable and unconventional sense of humor, which made every day at the institute utterly enjoyable. Thanks to Dr. Eur-Ing. Stylianos Rafailidis, Ms. Annette Pelke, Ms. Ulrike Putzke, Mr. Günther Garbe and Mr. Ralph Gehlsen-Lorenzen for their constant kindness and readiness to help. Also thanks to Tobias Becke for his feedback after reading the draft of this thesis and to Oliver Schülting for his great work in the development of the TransiEnt model library, back then as a working student and more recently as a researcher at the IET.

Thanks to friends from near and far, for being there when I need to take a break from work. Thanks to my family for their support and their teachings: to my parents for being great friends and role models in all stages of life, to my sisters for staying close despite the distance and to my nieces and nephews for representing the future we are all working for. Finally, my most sincere gratitude to my beloved wife Paula for her unconditional support and love. For the laughter, the music and the science. For the years we've lived together and the years that are still to come.

Hamburg, in July 2017
Ricardo Peniche Garcia

Acknowledgements

TABLE OF CONTENTS

ABBREVIATIONS AND SYMBOLS

Abbreviations

CHP	Combined Heat and Power
DHG	District Heating Grid
DSO	Distribution System Operator
EEG	Renewable Energy Sources Act
EG	Electric Grid
EnWG	Energy Industry Act
CCGT	Combined Cycle Power Plants
GG	Gas Grid
GHI	Global Horizontal Irradiance
GT	Gas Turbine
HRSG	Heat Recovery Steam Generator
IO	Input / Output
O&M	Operation and Maintenance
PEM	Polymer Electrolyte Membrane
PtG	Power to Gas
PtGtCHP	Power to Gas to Combined Heat and Power
PtH	Power to Heat
PV	Photovoltaic
RE	Renewable Energy
RE share DH	Share of RE in District Heating Consumption
RH	Renewable Hydrogen
ST	Steam Turbine
STOR	Storage
TSO	Transmission System Operator

Chemical Symbols

CO_2	Carbon dioxide
H_2	Hydrogen
KOH	Potassium hydroxide

Latin Symbols

c	Specific costs	€/MW, €/MWh, €/m, €/m^3
e	Specific CO_2 emissions	kg/MWh
E	Electric energy	MWh
HHV_{H2}	Higher heating value of hydro-	MJ/kg
LHV_{H2}	Lower heating value of hydro-	MJ/kg
$\dot{m}$	Mass flow	kg/s
M	Mass	kg
P	Electric power	MW_{el}
$\dot{Q}$	Heat flow	MW_{th}
Q	Heat	MJ
T	Temperature	°C
t	Time	s, h

Greek

η	Efficiency
σ	CHP coefficient
ζ	Fuel utilization efficiency

Indices

cons	Consumption
el	Electrical
ELY	Electrolyzer
fuel	fuel
H_2	Hydrogen
HP	Heating Plant
in	Input
out	Output
PtP	Power to Power
PtH	Power to Heat
PtCHP	Power to Combined Heat and Power
th	Thermal

LIST OF FIGURES

List of Tables

List of Tables

1 INTRODUCTION

1.1 Motivation

Since the end of the twentieth century, several countries have established mechanisms to promote the installation of renewable energy (RE) generation capacity. In particular, Germany has witnessed a significant increase in its RE capacity thanks to the feed-in-tariffs defined in its Renewable Energy Sources Act (in German: EEG) [1]. In 2016, the installed capacity of PV, onshore and offshore wind energy in Germany reached 40.8 GW_{el}, 45.5 GW_{el} and 4.1 GW_{el} respectively, leading to a total of 90.4 GW_{el} [2]. As a comparison, Germany's highest annual electric power demand is currently around 80 GW_{el}. The RE generation capacity is expected to increase further in the following years.

High shares of RE in energy systems usually lead to two types of balancing issues. The first issue is related to the fluctuating nature of the RE production profiles. The power production profiles of PV and wind power plants depend mainly on two variable meteorological parameters: global horizontal irradiance (GHI) and wind speed. Therefore, the production profiles of these RE plants cannot be scheduled to meet a given demand. If, for instance, the power output from RE is weak at a moment of high demand, then conventional power production units will have to supply the residual load. Residual load is defined as the difference between the net electricity demand and the RE production. On the other hand, if the RE power output is available at moments with low demand, it is usually curtailed.

The second issue is related to the regional clustering of RE plants. These plants are usually concentrated in regions with favorable meteorological parameters, such as high GHI and wind speed values. In Germany, this fact has led to a concentration of large amounts of onshore and offshore wind energy in the northern states. The transmission capacity to transport the RE production from generators in the north to demand centers in the south is currently insufficient. Although there are national plans for expanding the transmission capacities, these measures are costly and require several years to be completed. Currently, the excess power is either transmitted through the electric grids of neighboring countries or curtailed by the transmission system operators (TSO) by means of re-dispatch measures.

Cities and metropolitan regions have potential to integrate RE in their energy systems and contribute to solving the previously described issues. The German city of Hamburg is a good example of the potential of cities in the integration of renewable energies. Hamburg has significant electricity demand and consumption. With an annual electricity consumption of 12.9 TWh_{el} and a peak demand of around 2 GW_{el} in the year 2012, Hamburg accounts for around 2 % of the German total consumption and peak demand. Although the RE capacity within the city boundaries is small relative to the city's demand, the city's neighboring states allocate large RE production capacities, especially onshore and offshore wind energy. Furthermore, Hamburg has electricity and natural gas distribution grids as well as a large district heating grid. This infrastructure could facilitate the integration of RE by means of power-to-heat (PtH), power-to-gas (PtG) and central storage technologies. Finally, Hamburg has a conventional power park for the local production of electricity and heat, which could meet the net demand at moments with low RE production. For these reasons, Hamburg is chosen in this work as a case study to evaluate the integration of RE in urban energy systems.

To achieve this, the system can be regarded as an energy island capable of satisfying its own energy demands by minimizing the exchange of energy flows outside of the system boundaries. If this kind of regional balancing is replicated in other metropolitan regions, the curtailment of RE, the unplanned power flows through neighboring transmission systems, and the need of additional transmission lines could be reduced. In other words, the balancing of RE within self-sufficient metropolitan regions could lessen the stresses in the national transmission grid, in this case, the German transmission system. Finally, applying similar balancing methods at a national level could contribute to reduce future transmission bottlenecks in the continental energy system, in this case the European transmission system. This is the approach followed in this work.

1.2 Aim and scope

The goal of the present work is to analyze variations to Hamburg's centralized energy park and its operation strategy in order to find ways to increase the share of RE in the city's electricity and heat consumption, while reducing the exchange of energy flows through the system boundaries and ensuring the security of supply.

The reference system of this study is the energy system of the Hamburg metropolitan region in the year 2012. In addition, eleven system variations are defined in which an increase in the installed RE capacity is assumed for a representative year in the future (2050 or beyond). Each of the analyzed system variations are individually modelled and simulated for this representative year. The demand and potential RE production profile are the same for all variations. After simulation, a set of energetic, economic and environmental performance indicators allow the analysis and comparison of the system variations. These performance indicators include share of RE and combined heat and power (CHP) in electricity consumption, total annual costs, total annual CO_2 emissions and share of RE in district heating consumption.

The variations to the centralized energy park analyzed in this work include fossil power and combined heat and power plants with and without hydrogen co-firing, PtH and PtG units as well as hydrogen storage. Centralized variations to the energy park are favored because they usually have a better economic performance than decentralized schemes. This is, among other reasons, because large scale systems usually have lower specific investment costs and better efficiencies.

The described variations to Hamburg's central energy park were conceived in order to define a so-called "central oriented scenario". This and other three scenarios are the subject of research in a larger joint research project [3]. The other scenarios analyzed within the joint research project are focused mainly on distributed generation, demand side management as well as gas production and storage solutions. Some of the sub-system models used in this work were created by the project's programming development team.

1.3 Outline

The technologies implemented in the system variations in order to improve the integration of RE are presented in Chapter 2. These technologies include conventional power and CHP plants with hydrogen co-firing, electrode boilers, electrolyzers, as well as hydrogen storage. Besides, this chapter also addresses balancing issues of RE in different countries and gives an overview of studies addressing these issues with different approaches.

In Chapter 3, the reference energy system and the system variations are described in detail. The description includes the definition of the system boundaries and the composition of the energy park. The common dataset used in all system variations is also presented in this chapter. This includes electricity and heat demand profiles as well as RE production profiles.

The modeling of the energy system variations and its constituting subcomponents is the main topic of Chapter 4. The subcomponents include the production, storage and conversion units. The implementation of the selected operation strategies is also described in this chapter.

The simulation results are analyzed in Chapter 5 and Chapter 6. These include energy flow profiles as well as annual energy, costs and CO_2 emissions balances. The system variations which achieve a better trade-off between costs, CO_2 emissions and share of RE in the system's energy consumption are identified and discussed. Final remarks on this work are included in Chapter 7

2 CURRENT SITUATION, RESEARCH AND TECHNOLOGIES

2.1 RE integration issues in regional energy systems

Issues related to the regional balancing of electricity production and demand are common in countries with large variable RE production capacities. To illustrate this point, balancing issues in China, Denmark and Germany are presented in this section.

According to [4], the government of China requires that all utilities should have 8 % of RE capacity in their electricity production portfolio by the year 2020. But the installed wind power production represents already today a challenge for some of the country's inter-provincial transmission grids. Besides, most part of the wind parks in China are located mainly in the north of the country, whereas the hydro plants, which could be used to balance the fluctuations of RE, are mainly located in the country's south and center. Because of the lack of sufficient transmission lines, curtailment has led to a reduction in the wind energy's capacity factor despite an increase in installed capacity. An interesting fact about the Chinese case is that, according to [4] and [5], controlled trials of electricity dispatch schemes in which RE have feed-in priority (such as the German scheme) led to conflicts between RE and CHP plants, which are the primary heating source in Northern China.

Another example of regional balancing issues is offered by Denmark's "Great Belt Link". According to [6], Denmark's wind parks are mainly located in the country's west side (Jutland and Funen), which is synchronized with the central European grid, whereas the country's east side (Zealand) is synchronized with the Nordic grid, i.e. with Finland, Sweden and Norway. Because of this, both regions participate in different electricity markets: DK1 for West Denmark and DK2 for East Denmark. The electricity price in the DK1 market has been traditionally lower than the price in the DK2 market. This is, to a large extent, due to the high wind penetration and the merit order effect [7]. RE integration strategies, such as the use of electric boilers for district heating, flourished in Western Denmark mainly due to these favorable electricity prices. However, on the 26th of August of 2010 both regions got interconnected via the "Great Belt Link", a 600 MW High Voltage Direct Current transmission line. This increase in the available transmission capacity led to an increase in prices at the western side (see [8] for additional information regarding this effect). With higher electricity prices, the economic feasibility of RE integration technologies, such as electric boilers, has been significantly reduced. It has been estimated, that operating hours of these plants were reduced by 77 %, from around 1,772 h to 400 h [6]. This example shows that the regional RE integration approach can be in direct conflict with the approach of increasing the transmission capacity.

Finally, the regional balancing issues in Germany are addressed. Two evident technical consequences of these issues are the increasing interventions of the TSO by means of feed-in management and congestion management measures. The feed-in management (in German *Einspeisemanagement*) is defined in § 14 of the EEG and allows the TSO to curtail RE's power output to avoid situations in the electric grid which could jeopardize the security of supply. The congestion management (in German *Engpassmanagement*) is defined in § 13 of the German Energy Industry Act (EnWG) [9] and allows the TSO to interfere in the scheduled power production and trade to avoid situations in which the grid could become unstable. Re-dispatch and countertrading are two of these congestion management measures.

A recent example of how these interventions are becoming an increasing concern for the German TSO took place between January 9th and 11th of 2015 [10, 11]. In this weekend, the storm fronts "Elon" and "Felix" led to wind speeds of up to 160 km/h, leading to very high wind power outputs. It was reported that during this period, the wind power production was up to 30.7 GW_{el}. The geographic and timely imbalance of production and demand forced the TSO TenneT to intervene with re-dispatch measures accounting for 4.8 GW_{el}, leading to costs of €6 million. The TSO 50Hertz reported interventions on 6.7 GW_{el}, resulting in costs of around €7 million. This led to total costs of at least €13 million for the mentioned weekend. According to the German law, the costs derived from these measures are then charged to the consumers by means of increased grid fees. Similar situations were reported by the TSO in Winter 2011/2012 [12].

Another balancing issue in Germany which is worth mentioning is related to the so-called "dark doldrums" (in German: *Dunkelflaute*). As reported by [13] and [14], certain metrological conditions can lead to several days of low wind speeds and low solar irradiance. Additionally, low temperatures lead to higher electricity demand. Such an event occurred in the days around the 24th of January of 2017, which led to the fact that around 90 % of Germany's power demand had to be supplied by conventional power plants, such as coal, gas and nuclear power plants. These extreme examples put in evidence the issues related to increasing RE capacities in regional energy systems.

2.2 Related research studies dealing with RE integration

Over the last years, several studies have addressed technical and economic questions regarding the integration of RE in international, national and regional energy systems. Table 1 provides an overview of selected RE integration studies sorted according to their geographic scope.

Table 1 – Overview of related studies

Short name	Reference	Geographic scope	Analyzed region	Analyzed technologies	Time resolution	Simulation tool
EWIS	[15]	International	Europe	EG	1 h and ms	EWIS market and grid model, Netomac
SteinEtal	[16]	International	Europe	EG, STOR	1h	GAMS/CPLEX
DENA II	[17]	National	Germany	EG, STOR	1 h	DIME, DIANA (EWI-Köln)
WWSIS	[18, 19]	National	USA (West)	EG	1 h	GE-MAPS
EREIS	[20]	National	USA (East)	EG	1 h	GE-MARS, PROMOD IV
HK	[21]	Regional	Hong Kong	EG, DHG	1h	EnergyPLAN
AUGS	[22]	Regional	Augsburg	EG, DHG, GG	1h	URBS (C++ Tool)
BERL	[23]	Regional	Berlin	EG, DHG	1h	deeco
SYMBIOSE	[24]	Regional	Wien	EG, DHG, GG	n. a.	PSS SINCAL. MATLAB.
InsHHWilh	[25]	Regional	Hamburg	EG, DHG, STOR	1h	RESSI
MorbHH	[26]	Regional	Hamburg	Power plant, CHP, electricity market	1 h	deeco and deeco-s
MastPlaHH	[27]	Regional	Hamburg	Cumulative balances: el. energy, heat, and transportation	n. a.	n. a.
ExpHamb	[28]	Regional	Hamburg	DHG, STOR, renewable heat sources	1 h	BET-SysMod

** EG: Electric Grid, DHG: District heating grid, GG: Gas Grid, STOR: Storage*

International studies, such as EWIS [15] and SteinEtal [16] aim to identify power transmission challenges imposed by high RE in electric grids in very large areas, such as the European Union. These studies take a rather holistic approach which sometimes includes the modeling of electricity markets and physical power flows. In [15], a market model calculates dispatching schedules for the energy park of several market areas for one year. Each market area is considered as a "copper plate", i.e. internal transmission bottlenecks are neglected. Only cross-border transmission limitations between the market areas are considered by means of Net Transport Capacities. Additionally, steady-state load flow simulations of selected operation points with high regional wind outputs and low demand are conducted with a detailed grid model. This model considers national transmission capacities. Finally, dynamic simulations of disturbances such as frequency and voltage drops due to short circuits are conducted to analyze the transient stability of the system with a time resolution of milliseconds.

National studies, such as DENA II [17], WWSIS [18, 19] and EREIS [20] usually divide the country into regions, assume production and storage capacities and then perform energy balances on each region, considering export and import exchanges and transmission limitations between these regions. In some cases, the exchange with neighboring countries is also considered. In studies about Germany, the selected regions often correspond to the ENTSO-E identification system for the German TSO zones, as in [17]. In some other cases, the regions are defined according to the states' political boundaries accompanied with a detailed transmission grid topology, as in [29]. As an example of national studies, the DENA II study [17] aims to quantify the required upgrades in the transmission grid considering the RE production expansion until 2020. Non-transportable power is identified with the help of electricity market models and a power transfer model. Different methods for the integration of the calculated non-transportable power are analyzed, such as the expansion of the transmission lines or the storage of non-transportable power. The study delivers an estimate of the required additional transmission lines and makes recommendations for further studies.

Regional studies such as HK [21], AUGS [22], BERL [23] or SYMBIOSE [24] present diverse approaches and focus on different aspects of regional energy systems. Most studies conduct yearly analyses with hourly resolution. The RE production profiles are either given as data sets into the model or calculated within the model based on regional weather data. Some studies analyze only the electric grid, but others analyze the heating sector as well. A few studies such as [22] and [24] consider the electricity, the heating and the gas sectors. An additional distinction between these studies is the representation of the power and heat production units: several studies apply an aggregated approach, which means that production units of one particular type are represented with only one "aggregated" unit [21], while a few studies consider each unit individually. Some studies use linear optimization models with costs minimization as target function to determine the dispatch schedules of the different production units. Other studies favor a technical optimization approach, in which meeting the heat demand or both electricity and heat demand is done based on a fix priority system.

Table 1 also includes related studies conducted specifically for the city of Hamburg. In InsHHWilh [25], the integration of RE in Hamburg's Wilhelmsburg quarter is analyzed. This study considers electricity and heat demand profiles as well as production profiles of RE within the Wilhelmsburg quarter. The use of thermal storage, lead batteries for electricity storage, and the conversion from power to heat via heat pumps are investigated. The heat and electricity distribution grids are not modelled in this study. These are considered as ideal sources in moments when the local production via RE or storage is insufficient. In MorbHH [26], the electricity market model deeco-s is used to evaluate the potential economic viability of the newly built, coal-fired

1,654 MW_{el} power plant Moorburg, located in the south of Hamburg. The study compares this plant with an alternative gas-fired combined-cycle heat and power plant. Under the assumption of CO_2 certificate prices of 30 to 70 EUR/t CO_2, the study concludes that plants like Moorburg are not economically feasible, even with co-generation. In MastPlaHH [27], Hamburg's "master plan" for the achievement of the city's climate goals [30] is evaluated. This report is mainly based on statistical figures, such as yearly electricity or heat consumption. Its main goal is the quantification of the CO_2 emissions of the city and the evaluation of the effects that certain measures could have on these. This study favors the so-called consumption-based analysis (in German *Verursacherbilanz*) by which yearly energy consumptions are multiplied by average specific CO_2 emission factors. CO_2 emissions are assigned to the end users, categorized in industrial, transportation and residential sectors. The alternative to this type of analysis is the so-called production-based accounting (in German *Quellenbilanz*) in which the balancing calculations consider the CO_2 emissions of units located in the analyzed region. The study ExpHamb [28], commissioned by the city of Hamburg in 2014 and published in 2015, discusses alternatives to the coal-fired CHP plant in Wedel, located at the west site of the city's district heating grid. The simulation program BET-SysMod is used, which consists of a mixed-integer linear optimization model with hourly resolution. The study analyzes fossil and renewable alternatives to supply Hamburg's district heating grid and recommends the use of renewable heat sources and industrial heat sources, as well as gas-fired CHP units at Hamburg's west site. Certain similarities can be found between the ExpHamb study and the present work, but there are differences regarding their approaches, models, boundary conditions, performance indicators and conclusions. Other studies and projects related to Hamburg's energy system are summarized in [31], but are not further discussed in this work.

Based on the literature review presented above, it can be concluded that there seems to be no studies following this work's approach for the non-aggregated, dynamic analysis of central-oriented RE integration options in urban energy systems, under consideration of the system's total electricity and heat consumption.

2.3 Central-oriented technologies for RE integration

2.3.1 Conventional power and heat plants

Conventional power plants are essential in the successful integration of RE, because they have the ability to cover the residual load. Conventional power plant technologies considered in this work include steam power plants and combined cycle power plants (CCGT). The option of extracting heat from these plants for district heating purposes, i.e. co-generation, is also considered. These technologies are briefly described in this section. The relevant parameters of these representative conventional production units are summarized in Table 2.

As a convention, all efficiency values in this work refer to the net nominal production capacity of the plant, unless part-load efficiency considerations are explicitly stated. Besides, all efficiencies refer to instantaneous values, unless explicitly stated as annual efficiencies.

Table 2 – Relevant parameters of conventional power plants

	Maximum power of a single unit	**Electrical efficiency**	**Minimal load [% P_{nom}]**	**Ramp rate [%/min] @ % P_{nom}**	**Startup time [h] Hot (t_{off} < 8 h) Warm (t_{off} < 48 h) Cold (t_{off} <120 h)**
Steam power plant (hard coal)	750 – 1,600	45 – 47	20 – 25 (recirculation steam generator)	3 – 6 @ 40 –100	1.3 – 1.5 3 – 5 5- 10
Combined-cycle power plant	110 – 578	58 – 61	15 – 25 (2GT+1ST plant)	4 – 9 @ 40 –100	0.5 – 1 1 – 1.5 2 – 3

Sources: [32], [33].

Steam power plants are based on the Rankine cycle and consist of a steam generator, steam turbines at different pressure levels, a condenser, pumps, and feed water heaters. The fuel used in these plants is usually coal. In this work, brown coal is not considered as a fuel in the future system variations, so that the term coal refers exclusively to hard coal. The power plant Moorburg, located in Hamburg, is a good example of modern coal-fired steam power plants. It has an electrical efficiency of 46.5 %, defined as

$$\eta_{el} = \frac{P_{el,PP,out}}{\dot{Q}_{fuel,PP,in}}, \quad (1)$$

where $P_{el,PP,out}$ is the net electric power output of the plant and $\dot{Q}_{fuel,PP,in}$ is the heat input from fuel required by the power plant to produce power. This power plant consists of two steam generators with a thermal output of 1,627 MW_{th} each, and two turbine-generator sets with a nominal power production capacity of 827 MW_{el} each. Typical minimum load for coal-fired steam power plants is around 25 %. Typical ramp rates, i.e. the maximum allowable power output increment per unit of time, is around 3 to 6 %/min [32].

As the name suggests, natural gas combined-cycle power plants generate electricity by the combination of the gas and steam cycles. The hot combustion gases leaving the gas turbine deliver their heat into the steam cycle by means of a series of heat exchangers within a Heat Recovery Steam Generator (HRSG). Because of this combination, this type of plants achieve very high efficiencies. For instance, the Irsching 5 block is a CCGT unit with a power output of 860 MW_{el} and electrical efficiency of 59.7 % [34]. It consists of two gas turbines and one steam turbine. The flexibility of these plants is limited by the steam cycle. Ramp rates of around 4 to 9 %/min are state of the art [32].

Gas-turbine power plants are based on the Joule cycle and usually use natural gas as fuel. These plants consist of an air compressor, a combustion chamber and an expansion gas turbine. The Relizane power plant in north Algeria is an example of this kind of plants. With a total net power output of 465 MW_{el}, this plant consists of three gas turbines and three turbo generators. The electrical efficiency of this kind of plants is currently around 40 %. With current ramp rates of around 12 %/min [35], gas-fired power plants have the highest flexibility of all fossil power plants.

Heating plants are also an important component in urban energy systems. They use the heat of combustion to produce hot water or steam. Typical fuels are waste or natural gas. The heat efficiency of a heating plant is defined as

$$\eta_{th} = \frac{\dot{Q}_{th,HP,out}}{\dot{Q}_{fuel,HP,in}}, \tag{2}$$

where $\dot{Q}_{th,HP,out}$ is the heat output and $\dot{Q}_{fuel,HP,in}$ is the heat input from fuel required by the heating plant to produce heat. Typical heat efficiency values are around 95 % [36], which is the value used in this work. Although these plants are usually not considered as technologies which enable RE integration in the system, the production of fuels with RE sources (see section 2.3.2) and their usage in heating plants would enable this integration.

Finally, the concept of co-generation is briefly described. Co-generation is the simultaneous production of electricity and heat in a single generating unit. Large power plants with co-generation are usually called combined heat and power (CHP) plants. Large scale CHP plants are mainly steam cycle or combined cycle plants. An example of a coal-fired steam cycle CHP plant is the plant Tiefstack in Hamburg, with a power output of 205 MW_{el} and thermal output of 285 MW_{th} [37]. The plant Niehl 3 in Cologne is an example of a gas-fired combined-cycle heat and power (CCGT-CHP) plant, with nominal power and heat production capacities of 452 MW_{el} and 265 MW_{th} [38]. Smaller CCGT-CHP plants also exist, such as the unit with power output of 138 MW_{el} and a thermal output of 140 MW_{th} located as well in Tiefstack, Hamburg [39].

The electrical efficiency of a CHP plant is defined as

$$\eta_{el,CHP} = \frac{P_{el,CHP,out}}{\dot{Q}_{fuel,CHP,in}}, \tag{3}$$

where $P_{el,CHP,out}$ is the net electric power output of the CHP plant and $\dot{Q}_{fuel,CHP,in}$ is the heat input from fuel required by the CHP plant to produce power and heat.

The heat efficiency of a CHP plant is defined as

$$\eta_{th,CHP} = \frac{\dot{Q}_{th,CHP,out}}{\dot{Q}_{fuel,CHP,in}}, \tag{4}$$

where $\dot{Q}_{th,CHP,out}$ is the heat output of the CHP plant and $\dot{Q}_{fuel,CHP,in}$ remains as defined in equation (3).

The fuel utilization efficiency of a CHP plant is defined as

$$\xi_{CHP} = \frac{P_{el,CHP,out} + \dot{Q}_{th,CHP,out}}{\dot{Q}_{fuel,CHP,in}}, \tag{5}$$

where the constituting terms in this equation remain as defined in equations (3) and (4).

In recent years, several studies have addressed the questions regarding the technical challenges imposed by increasing shares of RE in electricity consumption [19, 40, 41]. These studies mostly conclude that conventional power plants will be subject to larger and more frequent mechanical and thermal stresses. Suitable materials, new component designs (for instance thinner pipe walls for steam generators) and new operation modes (such as the use of predictive control strategies) are some of the measures that can contribute to the adaptation of conventional production units to the new market conditions.

2.3.2 Conversion and storage in other energy forms

Power to Heat

For decades, producing heating water with electric energy has been a technical taboo for the German energy community. However, with increasing shares of RE in the national energy park, this type of heating is starting to regain acceptance under the term of Power to Heat (PtH). The concept behind this term is that the electric power surpluses from RE plants could be utilized to produce heat, despite the inherent exergetic losses of this process. This loss is generally accepted because the alternative would be just to curtail the excessive RE output. Instead, PtH units allow the coupling of the heat and power sectors.

According to [42], there are two main business cases for PtH units. The first possibility is to activate PtH capacities when the electricity prices are low (for instance due to the merit order effect). The second possibility is to activate PtH capacities to offer negative balancing power.

There are two main PtH technologies commercially available: electric process heaters and electrode boilers [42]. Electric process heaters consist basically of several heating rods, contained within a shell, that directly heat water coming from the district heating return pipeline. Electrically, these heaters are usually connected to low voltage bus bars. The temperature is regulated via power electronics. On the other hand, electrode boilers are tank-like components with electrodes inside. In this case, water with high conductivity is heated with an electric current flowing from one electrode to the other. The temperature is regulated by the positioning of the electrodes in the water. A heat exchanger then transfers the heat to the district heating grid. Electrode boilers are usually connected to medium voltage bus bars.

The Power to Heat efficiency of such a unit is defined as

$$\eta_{th,PtH} = \frac{\dot{Q}_{th,PtH,out}}{P_{el,PtH,in}}, \tag{6}$$

where $\dot{Q}_{th,PtH,out}$ is the heat output of the PtH unit and $P_{el,PtH,in}$ is electric power required to produce it.

In Hamburg, an electric steam generator with a capacity of 45 MW_{th} is already installed (EK Karoline). Plans of installing new units with capacity of 2 × 25 MW_{th} with the option of installing additional 2 × 25 MW_{th} have been reported [39]. Further PtH examples can be found in Nuremberg (50 MW_{th}), Flensburg (30 MW_{th}) or Herne (60 MW_{th}) [42], [43].

Power to Gas

Hydrogen production via electrolysis is a mature technology which has been used for decades, mainly in the chemical industry. In the 1990s, electrolysis gained importance as a way of potentially generating hydrogen with RE to be used in hydrogen fueled-transportation. Several demonstration projects have been built since then. However, the recent concerns regarding regional balancing of RE have triggered new interest in electrolysis as a way of integrating excessive RE power. Power to Gas (PtG) is the term coined to describe this concept.

Electrolysis is the process of splitting water into hydrogen and oxygen with the help of electric energy. There are two main electrolysis technologies: alkaline electrolysis and PEM electrolysis. The main difference between these two technologies is the used electrolyte. Alkaline electrolysis uses a solution of water and potassium hydroxide

(KOH) as electrolyte, whereas in PEM electrolysis the electrolyte is a Polymer Electrolyte Membrane (PEM). Alkaline electrolysis is the most mature technology, with nominal power capacities around 3.4 MW_{el} per unit. Large arrays consisting of several units have reached capacities of up to 160 MW_{el}. [44]. On the other hand, PEM electrolysis is a comparatively young technology which currently has nominal power capacities around 1.0 and 2.2 MW_{el} per unit, but large arrays consisting of several units can also be used to achieve capacities in the multi-megawatt range [45–47].

The lower heating value of hydrogen (LHV_{H_2}) equals 120 MJ/kg and the higher heating value of hydrogen (HHV_{H_2}) equals 141.86 MJ/kg. The performance of the electrolyzer units is determined by the hydrogen production efficiency of the electrolyzer, defined as

$$\eta_{ELY} = \frac{\dot{m}_{H_2,ELY,out} \cdot LHV_{H_2}}{P_{el,ELY,in}}, \tag{7}$$

where $\dot{m}_{H_2,ELY,out}$ is the electrolyzer's hydrogen mass flow output, LHV_{H_2} is the lower heating value of hydrogen and $P_{el,ELY,in}$ is the electric power required by the electrolyzer to produce hydrogen. In this work, an electrolyzer efficiency of 70 % according to this definition is assumed, which considers future technology improvements [48].

An overview of the PtG pilot plants can be found in [49]. The description of three of these PtG pilot plants follows. The first chosen example is the PtG plant in Falkenhagen, Germany. This plant generates hydrogen via 2 MW_{el} of alkaline electrolyzers and feeds it directly into the neighboring natural gas transportation grid. The hydrogen production efficiency of the electrolyzer is reported to be 64 % based on the HHV_{H_2}[49], which converted to LHV_{H_2} leads to a value of 54 %. The second example is the PtG plant in Hamburg-Reitbrook, Germany. This plant generates hydrogen via a 1.5 MW_{el} PEM electrolyzer and feeds it directly into the city's natural gas grid. The hydrogen production efficiency of the electrolyzer is reported to be 68 % based on the HHV_{H_2} [50], which converted to LHV_{H_2} leads to a value of 58 %.

The third example is the PtG plant in Werlte, Germany. This plant generates hydrogen via 6.3 MW_{el} of alkaline electrolyzers and then converts this into methane via a methanation process. This synthetic methane is then injected into the grid as biogas [49]. The hydrogen production efficiency of the electrolyzer is reported to be 70 % based on the HHV_{H_2} [49], which converted to LHV_{H_2} leads to a value of 59.2 %. Based on the lower heating value of methane, the electricity-to-methane efficiency of this plant is reported to be 54 % [51], which represent the ratio of the synthetic methane mass flow output to the electric power input required by the electrolyzer. The difference between the hydrogen production efficiency of the electrolyzer and the electricity-to-methane efficiency is due to the heat losses in the exothermic methanation process. These losses can be reduced if this heat would be utilized in other processes of the plant which require heat [52]. PtG plants with methanation are capable of feeding gas in the grid unrestrictedly, which is not the case for hydrogen-only PtG plants. Natural gas grid regulations usually allow only small amounts of hydrogen in the grid. According to German regulations, applications such as natural gas-fired cars and some gas turbines currently allow hydrogen concentrations of only 1 to 2 vol% [53], but hydrogen concentrations in the single-digit range are not considered critical for less stringent applications.

Hydrogen storage

As previously discussed, the potential of hydrogen injection in the natural gas grid is limited. Hydrogen storage presents an alternative to directly feeding hydrogen into the grid. There are mainly three technologies available at the moment for storing gaseous hydrogen: compressed gas vessels, metal hydrides and underground storage.

Compressed gas vessels are pressurized tanks made out of metal or fiber reinforced polymers in which hydrogen is stored at pressures between 350 and 700 bar. The metal hydride hydrogen storage technology consists of tanks filled with metal hydrides which change their chemical composition by absorbing hydrogen. This allows storing high quantities of hydrogen in relatively small volumes. However, thermal energy is required to release the hydrogen from the hydrides, and there are still concerns regarding the lifetime loss of the metal hydrides due to high load-unload cycling [54].

Finally, in underground hydrogen storage, underground salt domes are used to store large amounts of hydrogen at pressures usually between 45 and 150 bar. Typical geometrical capacities of this type of caverns are around 70,000 and 580,000 m^3. Examples of these caverns can be found in Teesside, UK (3 x 70,000 m^3) and Moss Bluff, USA (566,000 m^3) [48]. Some disadvantages of underground hydrogen storage are long planning and implementation times and the fact that this type of storage requires specific geological formations (salt domes) which cannot be found everywhere. Despite these disadvantages, this technology is chosen for this work because the Hamburg metropolitan region has favorable geographic conditions and because the large storage capacities attainable with this technology are very appropriate for the central-oriented approach of this work.

2.3.3 Recovery of stored energy

Hydrogen obtained with renewable energy can be directly used in chemical processes, for transportation or in stationary applications for power and heat production.

Fuel cells are the technology mostly associated with the conversion of hydrogen to electric power in stationary applications. However, high investment costs and low nominal power production capacities reduce the economic feasibility of energy systems based on this technology. For this reason, alternative technologies for the conversion of hydrogen to power and heat seem necessary.

CCGT plants are considered in this work as one of the most promising technologies for integrating renewable hydrogen (RH) in urban energy systems through the injection of hydrogen into the combustion chamber of the plant's GT. In the Dow facility in Stade, Germany, hydrogen was co-fired together with natural gas in GT for several years [55]. Some studies dealing with the future integration of hydrogen in energy systems list further examples of hydrogen co-firing in GT and are already assuming that pure hydrogen GT will be commercially available in the near future [48], [56]. The technical issues of hydrogen co-firing in GT are currently being addressed by gas turbine suppliers and research groups [57–61]. It has been reported that 15 vol% co-firing of hydrogen in reheat GTs can be handled [58]. According to [48], GT with single burner combustors can burn synthetic fuels with around 50 vol% hydrogen content.

3 METHODOLOGY

As already mentioned in previous chapters, the energy system of the Hamburg metropolitan region is the subject of this investigation. A group of scenarios is defined in order to analyze technical, economic and environmental aspects of the integration of RE in future urban energy systems. Every variation shares the same system boundaries but has a different energy park composition and a different operation strategy. In this chapter, the energy system and the variations analyzed are described. The description includes the system boundaries, the energy park composition as well as the data sets used for the calculations.

3.1 System boundaries

The system boundaries determine which demand profiles and which generating units are considered in the balance. These boundaries are valid for all system variations. They do not correspond to Hamburg's administrative boundaries but are closely related to the city's energy distribution infrastructure presented in Figure 1.

The definition of the boundaries for the electric system is perhaps the most important of all. This does not mean that the heating and gas infrastructures are less important, but the technical issues derived from increasing shares of RE currently manifests mainly on the electricity grid. The system boundaries selected for the definition of the electricity demand are the three transformer stations which connect the transmission grid (at 380 kV) with the city's electricity distribution grid (at 110 kV), located in the north, the east and the south of Hamburg. By making this assumption, electricity demand data made publicly available by the city's distribution system operator (DSO) can be directly used. As already mentioned in section 1.1, the approach followed in this work implies considering the system as an energy island which is able to satisfy its own energy demands at all times. Electricity exports are avoided either by curtailing excessive RE production or by using switchable loads which allow sector coupling between different forms of energy such as power, heat and gas. In particular, the use of electrode boilers for heat production or electrolyzers for hydrogen production is analyzed. The heat sector is mainly focused on the district heating grid, but other heating sources of the city such as gas, oil and electricity are also considered. Regarding the district heating sector, the system boundary is defined by Hamburg's main district heating grid shown in Figure 1.

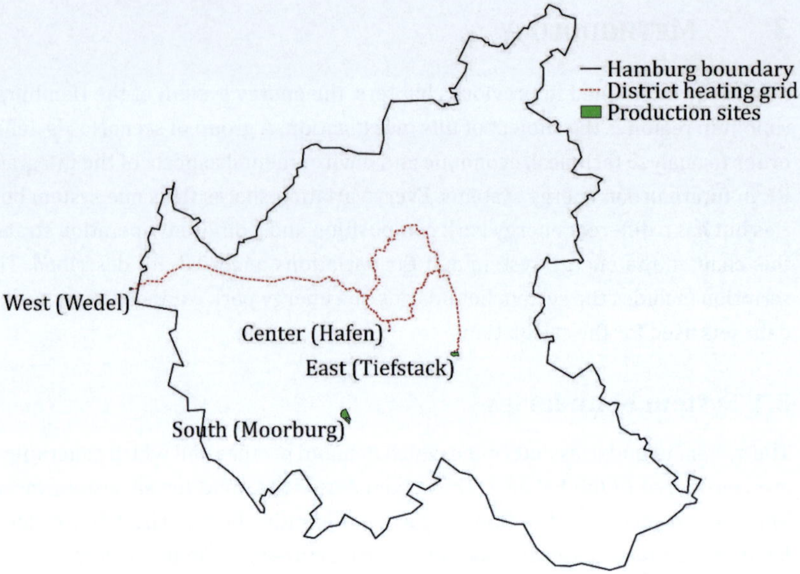

Figure 1 – System boundaries

The system boundary of the gas sector extends to Hamburg's gas distribution ring. In the present work, the gas grid is considered as an ideal source, i.e. capacity limitations and geographic resolution are not considered. Biogas fed into the grid is not considered. Hydrogen generated via PtG systems is not fed into the grid, but is first stored separately and then mixed with natural gas in a second step as described in section 4.2.3.

The four main production sites considered in this work are the East (Tiefstack), West (Wedel), Center (Hafen), and South (Moorburg) sites, as shown in Figure 1. Currently, coal-fired CHP plants are located at the East and West sites, a coal-fired power plant is located at the South site and a gas-fired heating plant is located at the Center site.

3.2 Demand profiles

The power and district heating demand profiles shown in Figure 2 are the starting point for the definition of all system variations. The power demand reported by Hamburg's DSO [62] corresponds to the year 2012 and has a time resolution of 15 minutes. It enables the observation of demand fluctuations produced by factors such as day and night, week-day or weekends, cold days and holydays. The maximum demand value reaches 2,011 MW_{el}, the minimum 917 MW_{el}. The yearly electric energy consumption equals 12.9 TWh_{el}.

The district heating demand was calculated with the help of a Hamburg-specific characteristic line published in [63]. This characteristic line defines a correlation between the ambient temperature and Hamburg's district heating production. In this work the terms district heating production and consumption are used interchangeably because the boundary defining the district heating consumption is considered to be at the limit between the heating producers and the grid so that heat losses in the grid are considered to be part of the total district heating consumption. The ambient temperature data provided by [64] was used as an input for the calculation of the load profile. A time resolution of one day (24 hours average) was selected for the ambient temperature. Although this time resolution leads to a loss in precision, it allows a better visualization of the sector coupling effects in several system variations. Additionally to the above described procedure based on the characteristic line, a heating water standard load profile [65] was superimposed in periods in which the heating demand was below the demand obtained with the standard load profile itself. In this way, the heating water demand could be considered in summer days with low room-heating demand. This procedure results in a maximum demand value of 1,250 MW_{th}, a minimum value of around 150 MW_{th} and an annual district heating consumption in Hamburg's main district heating grid of 3.9 TWh_{th}.

Since a general analysis of Hamburg's energy system is desired, the sole consideration of the district heating demand becomes insufficient. Therefore, the remaining technologies involved in the heating market must be also accounted for. Figure 3 shows the assumed total heating load profile of Hamburg. A total annual heat consumption of 22.6 TWh_{th} is assumed, with maximum demand of 6,890 MW_{th} and minimum values around 588 MW_{th}. This and other technology-specific heating load profiles were calculated by scaling the district heating demand profile so that the annual heating market shares published in [66] could be maintained. According to this publication, the heating market shares in Hamburg are as follows: 56 % of the market is covered by gas, 13 % by heating oil, 17 % by the main district heating grid, 5 % by other district heating grids, 8 % by electric heaters and the remaining 1 % by other sources. The resulting heat demand profiles remain unchanged in all system variations.

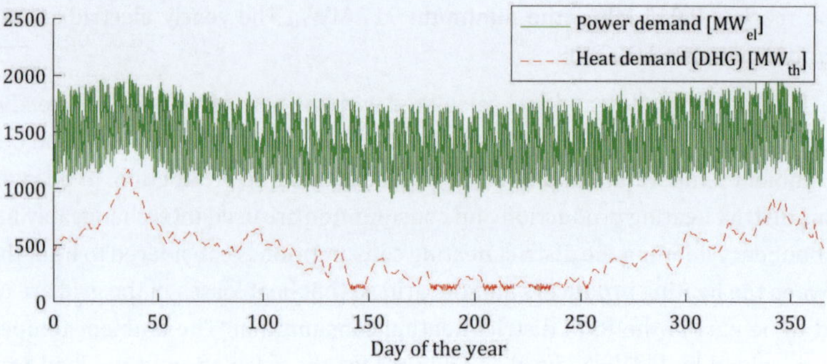

Figure 2 – Assumed electricity and district heating demand

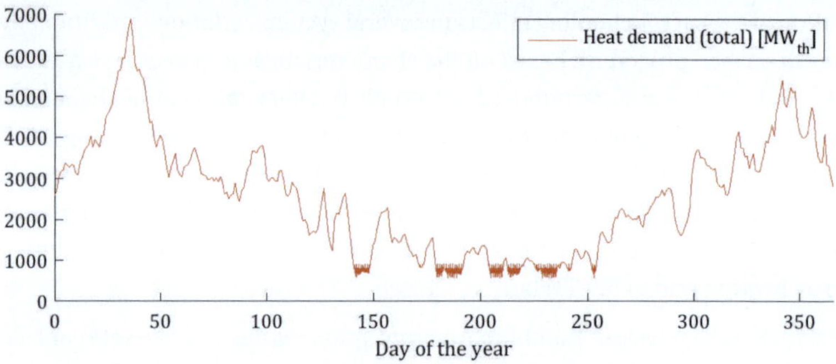

Figure 3 – Assumed total heat demand

3.3 Renewable energy contribution

Germany's RE installed capacity in the year 2012 as published in the TSO's national grid deployment plant [29] and the RE deployment targets published in the EEG were used as a starting point to define the RE contribution in the analyzed reference system and system variations in the future. The values obtained for Germany were then scaled down to the analyzed region with a scaling factor of 0.024. This factor was obtained by dividing Hamburg's annual electric energy consumption by Germany's consumption in the year 2012 (12.9 TWh_{el}/534.3 TWh_{el}). Then, the installed capacities of the year 2012 were adjusted so that the share of RE in electricity consumption would be similar to the value in Germany in the same year, which was 22.8 % according to [67]. The installed capacities of the future (referred to as the year 2050) were adjusted so that the RE potential would be almost equal to the electricity consumption of the analyzed system. The values resulting from these steps are shown in the columns containing the nominal power production capacity (P_{nom}) of each RE type for the years 2012 and 2050. The energy output (E_{gen}) of each RE type was obtained by scaling the production profiles reported by the TSO (50Hertz and TenneT) in the year 2012 with the P_{nom} factors of Table 3. The only exception to the usage of the published production profiles are the biomass plants, which are assumed to run continuously throughout the year to obtain the desired total output.

Table 3 - Installed RE capacities and resulting electricity production potential

	2012	2012	2050	2050
	P_{nom}[MW_{el}]	E_{out} [TWh_{el}]	P_{nom}[MW_{el}]	E_{out} [TWh_{el}]
PV	798	0.67	1624	1.35
Wind (onshore)	747	1.18	3044	4.78
Wind (offshore)	7	0.02	753	2.21
Run-off hydro	106	0.53	101	0.51
Biomass	100	0.57	390	3.41
Total	1758	2.97	5912	12.28

Figure 4 shows the resulting RE production profiles for the reference system (2012) and the rest of the system variations (2050) together with the system's power demand. It becomes clear that an increase in the installed RE capacity leads to a reduction in the residual load and to periods of time with RE surplus. The RE production profiles used in the simulation have a time resolution of 15 minutes.

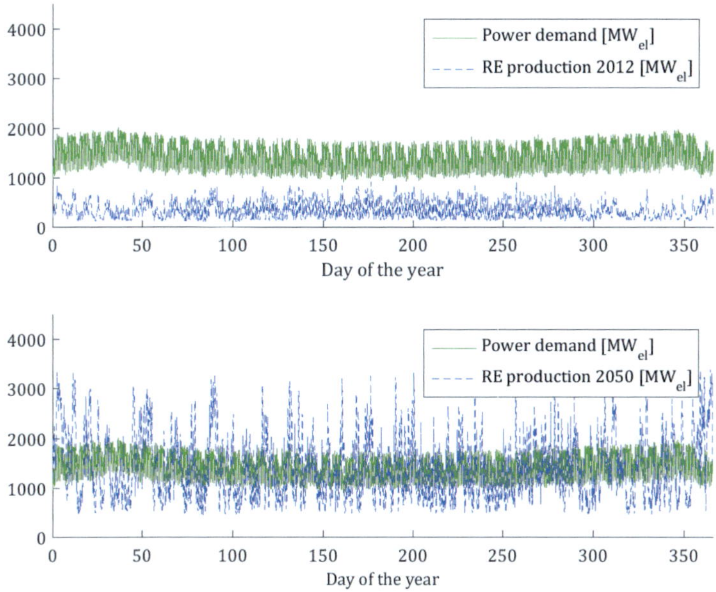

Figure 4 - Renewable energy production profiles for the years 2012 and 2050

3.4 Definition of system variations

As discussed in section 1.2, the effects of increasing the share of RE in electricity consumption on an energy park consisting of large scale power and CHP plants are analyzed in the present work. Besides, the effect of implementing central-oriented technologies for RE integration (see section 2.3) on the overall system is tested. Figure 5 shows an overview of the system variations analyzed in this work. Each variation includes a description of the changes with respect to the energy park of the preceding variation. The starting point is the reference system, which represents Hamburg's energy park with relatively low RE installed capacities corresponding to the year 2012 as described in the previous section. In all other system variations, a scaled RE installed capacity corresponding to the year 2050 is assumed (see Table 3). The variations can be classified in three main branches: A, B and C. In all the system variations of branch A (variations 1 to 6) a coal-fired power plant, consisting of two blocks, is considered at the south site. On the other hand, in the system variations of branch B (variations 7 to 9) it is assumed that one of the blocks at the south site is converted into a coal-fired CHP plant while the other block remains as a coal-fired power plant. Finally, the variations of branch C (variations 10 and 11) assume that no coal-fired plants are installed in the system.

In the reference system and variation 1, the south site (see Figure 1) is equipped with a coal-fired power plant consisting of two blocks with a nominal power production capacity of 827 MW_{el} each. At the east site (see Figure 1), a single coal-fired CHP block with maximum production capacity of 205 MW_{el} and 285 MW_{th} is considered. The power capacities of all CHP plants represent their maximum power production capacity in condensing mode and the maximum heat production capacity in backpressure mode and are based on characteristic diagrams published in [37]. Additionally, a waste-fired heating plant with nominal heat production capacity of 100 MW_{th} is assumed. This plant delivers its heat directly at the east site or via a heat converter located at the center site. Gas-fired auxiliary heaters for the peak heat demand are considered at the east site (320 MW_{th}) and the west site (25 MWth). At the west site, a coal-fired CHP plant consisting of two blocks is considered. The first block's maximum production capacity is 150 MW_{el} and 215 MW_{th} and the second block's maximum production capacity is 137 MW_{el} and 160 MW_{th}. The installed heat production capacity of the heating plant at the center site is 350 MW_{th} and it corresponds to the added capacity of three gas-fired hot water producers and two steam generators located in this heating station [36]. A pumped hydro storage similar to the one located in Geesthacht, Germany, with production capacity of 120 MW_{el}, pumping capacity of 96 MW_{el}, and 600 MWh of storage capacity is considered in the reference system. For the sake of clarity this plant is listed at the center site in Figure 5, since this component remains unchanged in all system variations.

In variation 2 the energy park considered in variation 1 remains unchanged, except for the units at the west site. There, the CHP plant is substituted by a gas-fired heating plant with a heat production capacity of 375 MW_{th}, which is equivalent to the heat production capacity of the coal-fired CHP blocks assumed at this site in the previous variations.

In variation 3 the energy park considered in variation 1 remains unchanged, except for the units at the west site. There, the coal-fired CHP plant is substituted by a gas-fired CCGT-CHP plant with a production capacity of 470 MW_{el} and 220 MW_{th} and a gas-fired auxiliary heater with a heat production capacity of 180 MW_{th}. The CCGT-CHP plant consists of two gas turbines and a single HRSG and steam turbine. Its configuration and nominal power production capacity are based on the descriptions found in an application for approval from the year 2012 [68]. The plant's nominal heat production capacity was obtained by conducting a thermodynamic simulation with the program EBSILON®Professional. The heat production capacity of the gas-fired auxiliary heater in this configuration is increased to 180 MW_{th} so that the total heat production capacity at the west site remains unchanged with respect to the reference system.

In variation 4 the energy park considered in variation 3 remains unchanged, except for the fact that a PtH unit with a nominal heat production capacity of 285 MW_{th} is considered at the east site. The heat production capacity of this unit equals that of the CHP plant at the east site. The heat from the PtH unit replaces the heat from the coal-fired CHP plant in periods of time with RE surplus.

In variation 5 the energy park considered in variation 3 remains unchanged, except for the fact that an electrolyzer park with a total power rating of 1,200 MW_{el} (30 × 40 MW_{el}) is considered. The hydrogen is produced by the electrolyzers with RE surplus and then stored in an underground cavern with a geometric capacity of 500,000 m^3. Hydrogen is co-fired in the gas turbine of the CCGT-CHP plant at a constant factor of 60 % of the plant's required heat input from fuel as long as hydrogen is available. In periods of time in which hydrogen is not available, natural gas provides all the plant's required heat input from fuel. In this variation, it is proposed that a cavern located at the Harsefeld salt dome is used for hydrogen storage and that a pipeline is installed to transport the hydrogen to the west site, where the CCGT-CHP plant is located, as shown in Figure 6. The estimated length of the hydrogen pipeline is 26 km. As a reference, the hydrogen grid in the Rhine-Ruhr metropolitan region [69] totals about 240 km. According to [70], the Harsefeld site already has caverns used for natural gas storage and retrofitting the existing storage capacity to allow hydrogen storage seems possible. Retrofitting a cavern might be less costs intensive than creating one from scratch. Still, it is assumed that every component of the energy park is a new installation, so that the full investment costs related to the cavern preparation are considered.

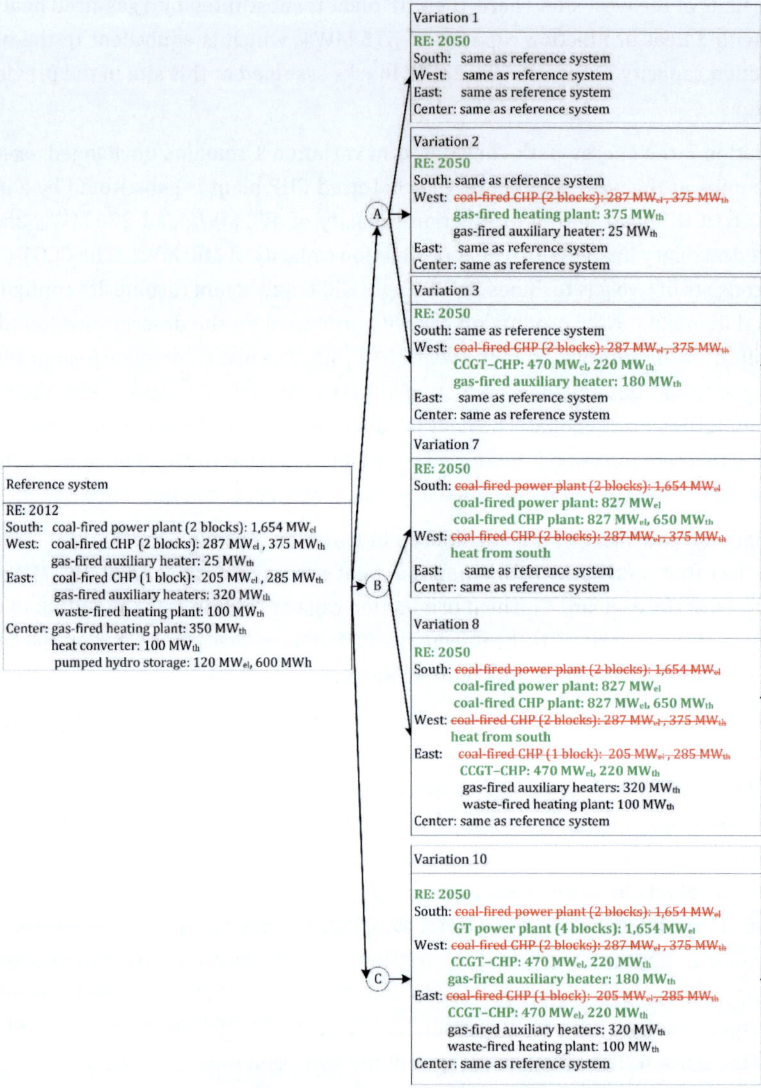

Figure 5 – Definition of system variations

Variation 4

RE: 2050
South: same as reference system
West: CCGT–CHP: 470 MW_{el}, 220 MW_{th}
gas-fired auxiliary heater: 180 MW_{th}
East: **additional Power to Heat: 285 MW_{th}**
coal-fired CHP (1 block): 205 MW_{el} , 285 MW_{th}
gas-fired auxiliary heaters: 320 MW_{th}
waste-fired heating plant: 100 MW_{th}
Center: same as reference system

Variation 5

RE: 2050
South: same as reference system
West: **Electrolyzers: 1,200 MW_{el}**
H_2 cavern storage: 500,000 m^3
CCGT–CHP: 470 MW_{el}, 220 MW_{th}
gas-fired auxiliary heater: 180 MW_{th}
East: same as reference system
Center: same as reference system

Variation 6

RE: 2050
South: same as reference system
West: **Electrolyzers: 1,200 MW_{el}**
H_2 cavern storage: 500,000 m^3
CCGT–CHP: 470 MW_{el}, 220 MW_{th}
gas-fired auxiliary heater: 180 MW_{th}
East: **additional Power to Heat: 285 MW_{th}**
coal-fired CHP (1 block): 205 MW_{el} , 285 MW_{th}
gas-fired auxiliary heaters: 320 MW_{th}
waste-fired heating plant: 100 MW_{th}
Center: same as reference system

Variation 9

RE: 2050
South: **additional Power to Heat: 650 MW_{th}**
coal-fired power plant: 827 MW_{el}
coal-fired CHP plant: 827 MW_{el}, 650 MW_{th}
West: heat from south
East: **Electrolyzers: 1,200 MW_{el}**
H_2 cavern storage: 500,000 m^3
CCGT–CHP: 470 MW_{el}, 220 MW_{th}
gas-fired auxiliary heaters: 320 MW_{th}
waste-fired heating plant: 100 MW_{th}
Center: same as reference system

Variation 11

RE: 2050
South: GT power plant (4 blocks): 1,654 MW_{el}
West: **Electrolyzers: 600 MW_{el}**
H_2 cavern storage: 250,000 m^3
CCGT–CHP: 470 MW_{el}, 220 MW_{th}
gas-fired auxiliary heater: 180 MW_{th}
East: **Electrolyzers: 600 MW_{el}**
H_2 cavern storage: 250,000 m^3
CCGT–CHP: 470 MW_{el}, 220 MW_{th}
gas-fired auxiliary heaters: 320 MW_{th}
waste-fired heating plant: 100 MW_{th}
Center: same as reference system

Variation 6 is a combination of variations 4 and 5. A PtH unit with a nominal heat production capacity of 285 MW_{th} is considered at the east site, which produces heat with otherwise curtailed RE, displacing the heat production of the coal-fired CHP plant. At the west site, a PtGtCHP plant is installed which consists of an electrolyzer park with a total power rating of 1,200 MW_{el}, an underground cavern with geometric capacity of 500,000 m^3 located at the Harsefeld salt dome, and a CCGT-CHP plant able to co-fire hydrogen at a constant factor of 60 % of the plant's required heat input from fuel as long as hydrogen is available.

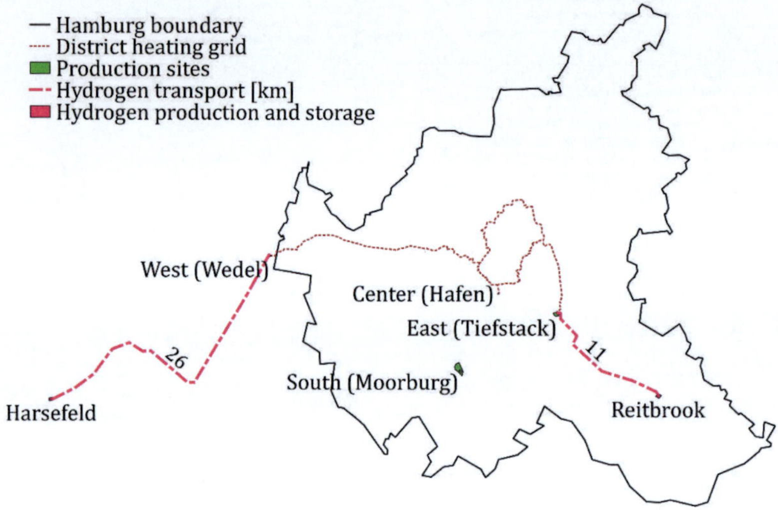

Figure 6 – Hydrogen production, storage, transport, and co-firing

In variation 7 a different approach is pursued which defines the branch B of the variation overview diagram shown in Figure 5. In this variation, it is assumed that one of the two coal-fired blocks at the south site is provided with co-generation capabilities. According to [71], the plant currently located at the south site could actually produce 650 MW_{th} of heat, and the plant's operation range goes from 35 to 100 %. In this variation, the CHP plant consists of two blocks with a total power production capacity of 1,654 MW_{el} and heat production capacity of 650 MW_{th}. It is assumed that one block is operated as a condensing power plant in charge of the residual load and the other block has co-generation capabilities. Additionally it is assumed that the coal-fired CHP plant at the west site is removed and its heat contribution is overtaken by the CHP plant located at the south site.

In variation 8 the energy park considered in variation 7 remains unchanged, except for the fact that the coal-fired CHP plant located at the east site is substituted with a gas-fired CCGT-CHP plant with production capacity of 470 MW_{el} and 220 MW_{th}.

Variation 9 is similar to variation 8 in most aspects, except for the fact that a PtH unit with a heat production rating of 650 MW_{th} is considered at the south site and a PtG unit with a total power rating of 1,200 MW_{el} and an underground salt cavern with geometric capacity of 500,000 m^3 is considered close to the east site for hydrogen storage. The dimensions of the hydrogen production and storage unit are equal to those considered in variation 5, but the location is different: in this case it is proposed that a salt cavern is built in the salt dome located in Hamburg-Reitbrook, as shown in Figure 6. A hydrogen distribution pipeline of 11 km length is assumed between Reitbrook and the east site, where the hydrogen is co-fired in the CCGT-CHP plant in such a way that 60 % of the heat input from fuel required by the plant is covered by hydrogen as long as there is hydrogen available. If hydrogen is not available, all the required heat input is provided by natural gas.

Variation 10 and variation 11 are part of branch C, which was created to evaluate the consequences of removing coal completely from the analyzed energy system. In variation 10, the coal-fired CHP plants at the east and west sites are substituted by gas-fired CCGT-CHP plants with nominal production capacities of 470 MW_{el} and 220 MW_{th}. At the south site, a GT power plant consisting of four blocks, each with a power production capacity of 413.5 MW_{el}, is considered.

Variation 11 is an extension of variation 10 but hydrogen co-firing is considered in the CCGT-CHP plants at the east and west sites. The Harsefeld and Reitbrook sites are provided with hydrogen production units, salt caverns for hydrogen storage and hydrogen transmission pipelines to transport the hydrogen to the CCGTs, as shown in Figure 6. The hydrogen production and storage capacities of both sites together equal the capacities assumed in variations 5 and 9. That means that the installed capacities at each of these sites is 600 MW_{el} and the cavern size is 250,000 m^3. No PtH unit is assumed in this variation due to its negative effect on the share of CHP power in electricity consumption, as explained in sections 5.5 and 6.3 (see definition of the "CHP curtailment effect").

3.5 Evaluation and analysis

Each of the system variations is first analyzed with respect to their time-dependent behavior by using weekly and yearly plots of the electricity and district heating production profiles. This allows the visualization of the consequences of changing the energy park's composition and operation strategy. This analysis of the production profiles is included in chapter 5.

In a second step, the system variations are analyzed with respect to their costs effectiveness, their environmental impact and their energy balance on a full-year basis. The costs effectiveness is evaluated by means of the system's total annual costs. The environmental impact is measured with the resulting total annual CO_2 emissions. The calculation of these quantities is described in detail in section 4.4. The energy balances of selected variations are illustrated by means of Sankey diagrams, which provide a good visualization of the calculated energy flows. This analysis of the annual performance indicators of the system variations is presented in chapter 6.

4 MODEL DESCRIPTION

4.1 Modeling of urban energy systems

Urban energy systems today consist mainly of producers, consumers and distribution grids. In this work, the analyzed system variations were modeled and simulated with the *TransiEnt* model library [3] developed in the programming language *Modelica* with the simulation environment *Dymola*. To illustrate the modelling approach, Figure 7 shows the graphical interface of a simplified exemplary system model.

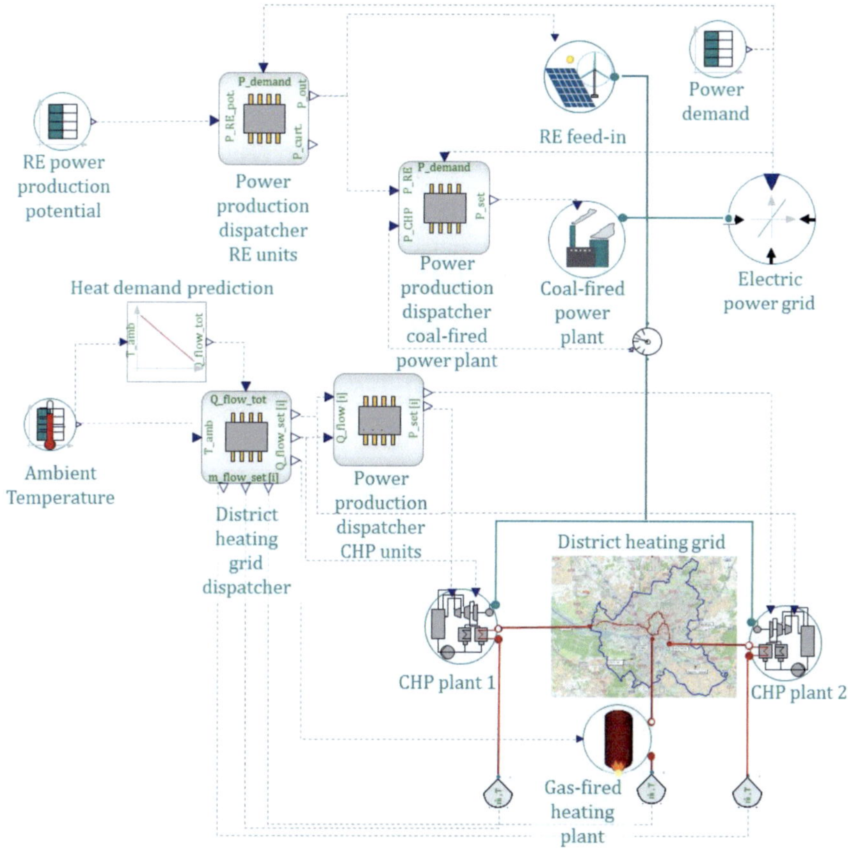

Figure 7 - Exemplary model topology

Regarding the power and heat production, a coal-fired power plant, two CHP plants, a heating plant and a component for the RE output can be identified. The power and the district heating grid are also depicted in this figure. The power grid is represented by a single node model and the district heating grid presents three feed-in points. Finally, four components are in charge of the superordinate determination of set points: the district heating dispatcher and the power dispatcher of the CHP units, the condensing power plant, and the RE units. The components shown in the figure contain in turn different systems of equations, characteristic lines, characteristic fields and interfaces, which enable their modeling with a level of detail adapted for annual simulations. The contents of the aforementioned components are described in the following sections.

4.2 Producer models

4.2.1 Power plants

The power plant model relies on three main parameters: the plant's nominal electrical efficiency, power output gradient and a relative part-load efficiency characteristic line. The part-load efficiency shown in Figure 8 is obtained by multiplying the plant's nominal electrical efficiency by the relative part-load efficiency characteristic line of the type of plant selected. The later were obtained from [72] and [73]. On the other hand, the dynamic behavior of the plant's power output is implemented by using linear time invariant (LTI) elements, a simple rule of proportion and the time constant definition, according to which the time constant of any system is equal to the time required for the observed variable to achieve 63.2 % of the set value.

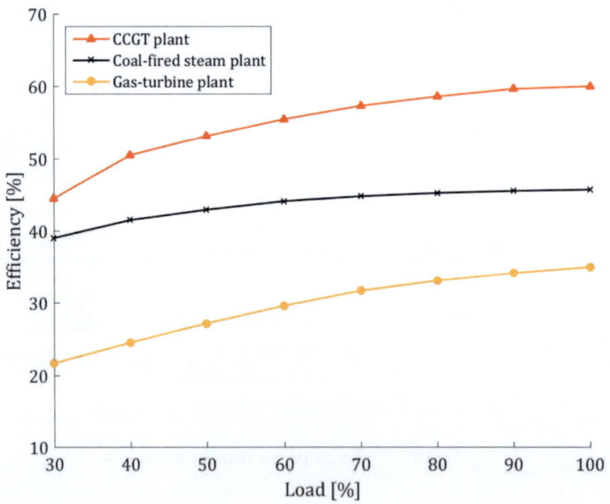

Figure 8 – Part-load efficiencies of the power plant types considered

4.2.2 Combined Heat and Power plants

All CHP plants in this work are considered to have a condensing-extraction turbine. The CHP plant model relies on three main parameters: the plant's operation limit characteristic lines (PQ diagram), the plants' heat input characteristic field, and the plant's power and heat output gradient.

Figure 9 shows the PQ diagram and the heat input characteristic field of a CHP plant. In this diagram, the plant's heat output is assigned to the horizontal axis, and the plant's power output is assigned to the vertical axis. The lines displayed therein define the operation limits of the plant. All possible operation points can be represented with a combination of power and heat output located within these operation limits. The upper limit is determined by the plant's full-load thermal input. The lower limit is determined by the plant's minimum part-load thermal input. The line connecting the upper and the lower limit is the backpressure line, which represents the maximum possible cogenerated power output for any given heat output within the plant's operation limits. Most of the PQ diagrams used in this work were obtained from [37], but some of them were calculated with the power plant simulation software EBSILON®Professional. Figure 9 also shows the heat input characteristic field of a CHP plant. The diagram shows the required heat input from fuel required by the plant for any given combination of power and heat output within the plant's operation limits. The fields used in the CHP models were calculated with the power plant simulation software mentioned before.

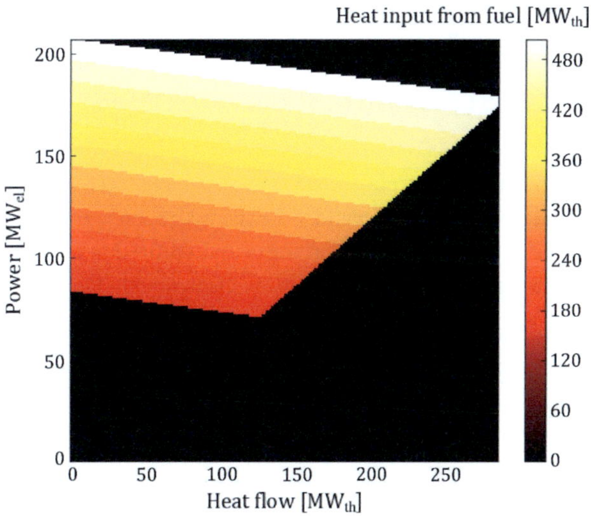

Figure 9 – PQ diagram and heat input characteristic field of a CHP plant

The dynamic behavior of the CHP plants was approximated with the simplified approach already described in the previous section, which consists of reproducing the plant's rated power and heat gradients by using LTI elements, the rated output gradients, a proportionality rule and the time constant definition.

In this work, the share of CHP power in electricity consumption (CHP share,el) is not the total electricity production of the CHP units. Only the electricity produced in cogeneration is considered. For the calculation of the share of CHP power in electricity consumption, the definition of the plant's CHP coefficient is used

$$\sigma_{CHP} = \frac{E_{el,CHP,out}}{Q_{th,CHP,out}}, \tag{8}$$

where $E_{el,CHP,out}$ is the plant's cogenerated electric energy output, and $Q_{th,CHP,out}$ is the heat output of the CHP plant.

The CHP power is therefore calculated by multiplying the time-dependent heat output of a CHP plant by its corresponding CHP coefficient. Typical CHP coefficients are provided in the European CHP directive (see the annex II in [74]). The CHP coefficients used in this work are shown in Table 4.

Table 4 – Selected CHP coefficients

Type of unit	Default CHP coefficient
Combined cycle gas turbine with heat recovery	0.95
Steam condensing extraction turbine	0.45

4.2.3 Power to Gas to Combined Heat and Power plants

In this work, Power to Gas to Combined Heat and Power (PtGtCHP) plants are energy conversion units in which RE surplus is converted into hydrogen which is then stored in underground salt caverns. In a second step, the hydrogen is transported via pipeline to a site where it is mixed with natural gas and co-fired in a CCGT-CHP plant to produce power and heat. This type of systems is illustrated in Figure 10.

The amount of hydrogen produced in the electrolyzers is obtained multiplying the electric power input by the hydrogen production efficiency of the electrolyzer (defined in section 2.3.2) and dividing the product by the lower-heating value of hydrogen. In terms of the equation (7) this can be expressed as follows

$$\dot{m}_{H_2,ELY,out} = \frac{\eta_{ELY} \cdot P_{el,ELY,in}}{LHV_{H_2}}. \tag{9}$$

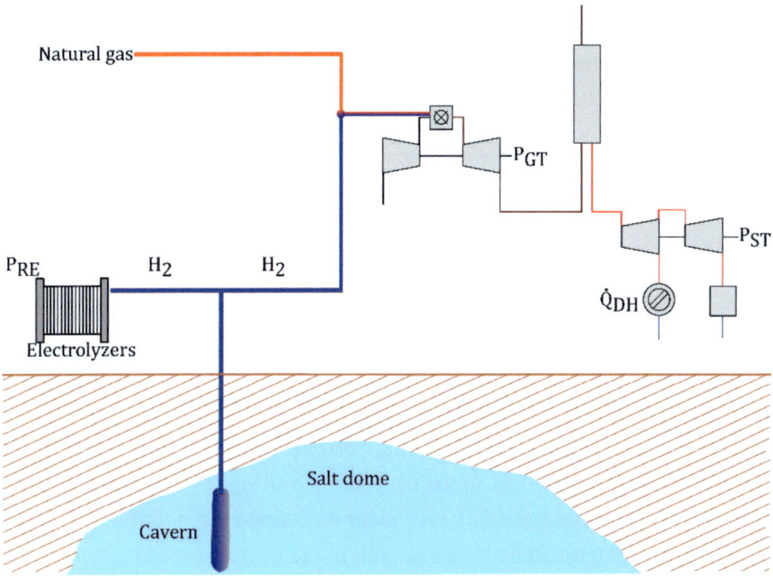

Figure 10 – Power to Gas to Combined Heat and Power (PtGtCHP) plant

The storage of hydrogen is modeled by a mass balance and a logic control algorithm. The hydrogen stored at a given time step is equal to the hydrogen produced by the electrolyzer minus the hydrogen sent to the CCGT-CHP unit. The equation involved is

$$\begin{aligned} \dot{m}_{H_2,stored} &= \dot{m}_{H_2,STOR,in} - \dot{m}_{H_2,STOR,out} \\ &= \dot{m}_{H_2,ELY,out} - \dot{m}_{H_2,CCGT\text{-}CHP,in}, \end{aligned} \tag{10}$$

where $\dot{m}_{H_2,stored}$ is the hydrogen mass flow being effectively stored in the cavern, $\dot{m}_{H_2,STOR,in}$ is the hydrogen mass flow sent into the cavern, $\dot{m}_{STOR,out}$ is the hydrogen mass flow extracted from the cavern, $\dot{m}_{H_2,ELY,out}$ remains as defined in (9), and $\dot{m}_{H_2,CCGT\text{-}CHP,in}$ is the hydrogen mass flow input of the CCGT-CHP plant.

The hydrogen mass flow input of the CCGT-CHP plant is obtained first by calculating the amount of hydrogen required to produce the desired heat input from hydrogen combustion in the CCGT-CHP plant and then by running a logic control algorithm which only allows the co-firing if hydrogen is available. The main equations involved are

$$\dot{Q}_{H_2,CCGT\text{-}CHP,in} = k_{co\text{-}firing} \cdot \dot{Q}_{fuel,CCGT\text{-}CHP,in}, \tag{11}$$

and

$$\dot{m}_{H_2,CCGT\text{-}CHP,in} = \frac{\dot{Q}_{H_2,CCGT\text{-}CHP,in}}{LHV_{H_2}}, \tag{12}$$

where $\dot{Q}_{H_2,CCGT\text{-}CHP,in}$ is the heat input from hydrogen combustion in the CCGT-CHP plant, $k_{co\text{-}firing}$ is the hydrogen co-firing rate, $\dot{Q}_{fuel,CCGT\text{-}CHP,in}$ is the heat input from fuel of the CCGT-CHP plant, and $\dot{m}_{H_2,CCGT\text{-}CHP,in}$ remains as defined in equation (10).

The fraction of the electricity and heat produced in CCGT-CHP plants by co-firing hydrogen produced with RE surplus in PtG units can be considered as renewable. This phenomenon is called the "RH heat and power contribution" in this work. From an annual perspective, this contribution is calculated with the following equations

$$E_{el,H_2,out} = \int_{t=0}^{T} P_{el,H_2,out} \cdot dt = \int_{t=0}^{T} \eta_{el,CCGT\text{-}CHP} \cdot \dot{Q}_{H_2,CCGT\text{-}CHP,in} \cdot dt, \qquad (13)$$

$$Q_{th,H_2,out} = \int_{t=0}^{T} \dot{Q}_{th,H_2,out} \cdot dt = \int_{t=0}^{T} \eta_{th,CCGT\text{-}CHP} \cdot \dot{Q}_{H_2,CCGT\text{-}CHP,in} \cdot dt, \qquad (14)$$

where $E_{el,H_2,out}$ and $Q_{th,H_2,out}$ are the electricity and heat produced with hydrogen from the PtG units, $P_{el,H_2,out}$ and $\dot{Q}_{th,H_2,out}$ are the power and heat outputs produced with hydrogen from the PtG units, $\eta_{el,CCGT\text{-}CHP}$ and $\eta_{th,CCGT\text{-}CHP}$ are the electric and thermal efficiencies of the CCGT-CHP plant as defined in equations (3) and (4), and $\dot{Q}_{H_2,CCGT\text{-}CHP,in}$ remains as defined in equation (11).

From these equations, the share of RE in the electricity and district heating consumption provided by the PtGtCHP plants can be calculated as follows

$$RE_{Share,el,PtGtCHP} = \frac{E_{el,H_2,out}}{E_{el,cons}}, \qquad (15)$$

$$RE_{Share,DH,PtGtCHP} = \frac{Q_{th,H_2,out}}{Q_{th,DH,cons}}, \qquad (16)$$

where $RE_{Share,el,PtGtCHP}$ is the share of RE in electricity consumption provided by the PtGtCHP plant, $E_{el,cons}$ is the system's electricity consumption (12.9 TWh_{el}), $RE_{Share,DHN,PtGtCHP}$ is the share of RE in district heating consumption provided by the PtGtCHP plant, $Q_{th,DH,cons}$ is the system's district heating consumption (3.9 TWh_{th}), and $E_{el,H_2,out}$ as well as $Q_{th,H_2,out}$ remain as defined in equations (13) and (14). In variations where PtGtCHP plants are present, these terms are considered for the calculation of the total shares of RE in electricity and district heating consumption (see section 4.4.3).

Besides the terms defined above, the definition of two efficiency sets is required for the evaluation of PtGtCHP plants with co-firing of hydrogen and natural gas. PtGtCHP plants have more than one product, and more than one energy source. In this work, a set of efficiency values for evaluating PtGtCHP plants is proposed. The first set is called the partial-input-output efficiency set (partial IO) for two reasons: a) it only takes into consideration the RE electricity input of the plant, dismissing its natural gas requirements; and b) it only considers the fraction of the electricity and heat outputs which can be attributed to the co-firing of RH. The second set is called the total input-output efficiency set (total IO) because it considers a) all energy inputs required by the PtGtCHP plant to function, and b) all its useful outputs.

The partial IO efficiency set consists of the PtP efficiency, the PtH efficiency and the PtCHP efficiency of the PtGtCHP plant. The total IO efficiency set consists of the electrical efficiency, the heat efficiency and the energy utilization efficiency of the PtGtCHP plant. These sets are used in section 6.2 to analyze the annual performance of PtGtCHP plants.

The PtP efficiency of a PtGtCHP plant is

$$\eta_{PtP,PtGtCHP} = \frac{P_{el,H_2,out}}{P_{el,ELY,in}}, \tag{17}$$

with the constituting terms in this equation as defined in equations (13) and (7).

The PtH efficiency of a PtGtCHP plant is defined as

$$\eta_{PtH,PtGtCHP} = \frac{\dot{Q}_{th,H_2,out}}{P_{el,ELY,in}}, \tag{18}$$

with the constituting terms in this equation as defined in equations (14) and (7).

The PtCHP efficiency of a PtGtCHP plant is defined as

$$\xi_{PtCHP,PtGtCHP} = \frac{P_{el,H_2,out} + \dot{Q}_{th,H_2,out}}{P_{el,ELY,in}}, \tag{19}$$

with the constituting terms in this equation as previously defined.

The electrical efficiency of a PtGtCHP plant is defined as

$$\eta_{el,PtGtCHP} = \frac{P_{el,CHP,out}}{P_{el,ELY,in} + \dot{Q}_{NG,in}}, \tag{20}$$

where $P_{el,CHP,out}$ and $P_{el,ELY,in}$ remain as defined in equations (3) and (7), and $\dot{Q}_{NG,in}$ is the heat input in form of natural gas required by the CHP unit to produce power and heat.

The thermal efficiency of a PtGtCHP plant is defined as

$$\eta_{th,PtGtCHP} = \frac{\dot{Q}_{th,CHP,out}}{P_{el,ELY,in} + \dot{Q}_{NG,in}}, \tag{21}$$

where $\dot{Q}_{CHP,out}$ remains as defined in equation (4) and the rest of the constituting terms remain as previously defined.

The energy utilization efficiency of the PtGtCHP plant is defined as

$$\xi_{PtGtCHP} = \frac{P_{el,CHP,out} + \dot{Q}_{CHP,out}}{P_{el,ELY,in} + \dot{Q}_{NG,in}}, \tag{22}$$

with the constituting terms in this equation as previously defined.

4.2.4 Power to Heat units

According to [42], PtH units are usually connected in parallel to other heat producers in the heating water system. In this work however, the PtH units are connected in series with the heating condenser of large scale CHP plants so that the heating water from the district heating's return first passes through the CHP plant's heating condenser and then through the PtH unit. To illustrate how this can lead to a higher electrical efficiency of the CHP plant, Figure 11 shows a schematic diagram of a CHP plant with two heating condensers connected in series with a PtH unit. In this example, the first heating condenser is supplied with steam extracted from the last stage of the middle pressure turbine (E2) and the second one is supplied with steam extracted from a previous stage (E1) with a slightly higher pressure level. In this configuration, when the PtH unit is active, the steam extraction E1 can be gradually closed and the steam which was previously extracted at E1 can now be sent through the next turbine stage. By doing so, the power output of the turbine can be kept constant while the heat input from fuel of the plant's steam generator is reduced, leading to an increase in the plant's electrical efficiency. Besides, the CHP plant is released from its heat production commitment and its power output can also be reduced. In this work, the PtH units are designed in such a way that they are able to cover the heat production of a CHP plant completely. When the PtH unit operates at its nominal operation point, no steam is extracted from the CHP plant.

In the model, the PtH unit receives a set value defining the amount of heat that the unit should produce. This set value is transferred directly to the heating water, since a Power to Heat efficiency of one (as defined in section 3.4) is assumed. The PtH controller operates in such a way that the unit supplies heat within its operational range as long as RE surplus is available. The set value of the CHP plant's heat production is reduced accordingly.

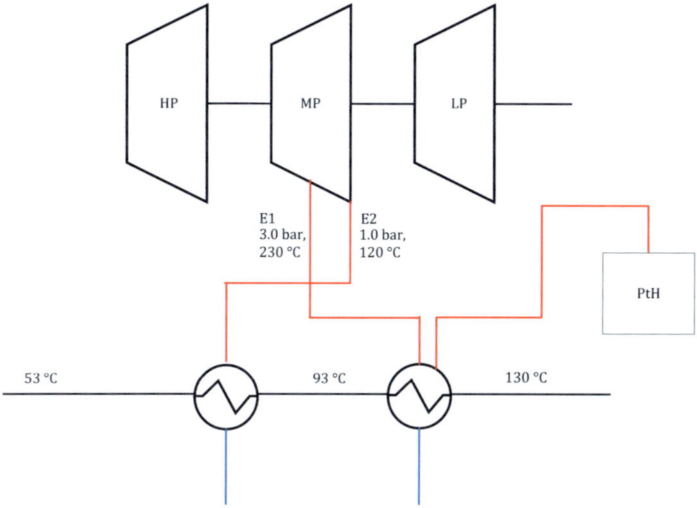

Figure 11 - PtH connection in series

4.3 Unit commitment and dispatch

In practice, the unit commitment problem is solved with mixed-integer linear optimization models, where the minimization of costs is the objective function. In this work, the set values required by the system components are provided by superordinate power and heat dispatchers implemented in Dymola itself. The optimization algorithm is substituted by logic controllers, which emulate a dispatch schedule based on a previously defined merit order: the production units are ranked in advance according to their fuel costs and electrical efficiency. Although the resulting dispatch schedule is not necessarily the solution with the minimum costs, it represents a reasonable approximation to common power market conditions today. A crucial advantage is that this approach reduces the amount of modeling tools required to simulate the analyzed energy system.

4.3.1 District heating grid dispatcher

The calculation of the heat demand profile of the main district heating grid was described in section 3.2. This demand must be covered by several production units. The district heating dispatcher determines how much of the district heating demand should be covered by which heat producing unit. There are three main elements which coordinate this task: the set values of the supply and return temperatures, the set values of the heating water mass flow at each feed-in point, and finally the load trimming between blocks located at the same feed-in point.

The set value of the supply temperature is a property of the district heating grid usually established in an early design stage and is usually in function of the ambient temperature of the supplied region. For periods of time in which the ambient temperature is relatively high, the supply temperature is set to the temperature level required by heat consumers that operate the whole year. For periods of time in which the ambient temperature is low, the supply temperature is set to a value at which the heating of buildings is guaranteed. In most cases, the set supply temperature increases proportionally with decreasing ambient temperature so that the required heating water mass flow, and therefore the power demand of the district heating pumps, are reduced. The set value of the return temperature is modeled by means of the so-called "return feed-in law", which is based on historic operation values. Both supply and return set values were obtained from [63].

The district heating grid model consists of four feed-in points at the east, west, center (Hafen) and center (Spaldingstr.) which deliver the heating water mass flow to a single mass flow sink. The set value of the heating water mass flow at each feed-in point of the main district heating grid was obtained by merging data from [63] and [75]. The result of this procedure is shown in Figure 12. The total amount of circulating heating water is then allocated to each of the four feed-in points of the analyzed district heating grid by means of the set value curves shown in this figure. These curves already consider the hydraulic restrictions of the grid. The time dependent value of the total heating water circulating in the heating grid is calculated by dividing the heat production set value of the whole grid by the enthalpy difference between supply and return throughout the year.

Finally, the set value trimming between blocks located at the same feed-in point is also modeled via characteristic lines. In the modeled energy system, this feature is relevant for modeling the west feed-in point, where the scheduling of two coal-fired CHP blocks is required. The set-value trimming strategy for these blocks was taken from [76]. The heat producers at the east feed-in point also require a set-value trimming strategy, which was modeled based on the values presented in [77].

Distribution pumps, pressure control schemes or transport pipelines are not depicted in the model. However, this simplification does not have a negative effect on the model's main performance indicators (e.g. fuel consumption, CO_2 emissions, CHP shares or RE shares) because the operation of the grid remains unaltered between system variations. Only the types of heat producing units at each feed-in point are modified.

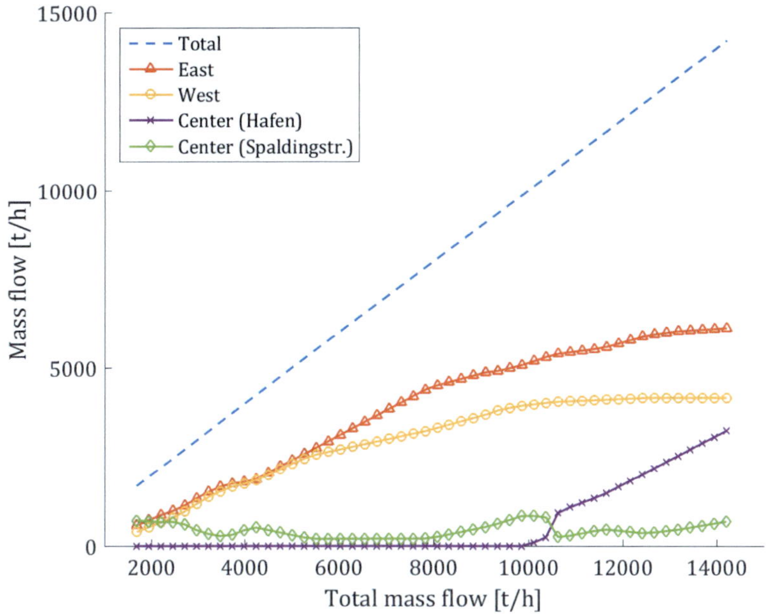

Figure 12 - Mass flow set values

4.3.2 Power production dispatcher

In the energy system model, the power production set value of all active producers is defined by a central control unit. The power production dispatcher is in charge of assigning an appropriate operation point for each producer under consideration of the total power load and each plant's design parameters and operational limits. The logic structure of such power production dispatchers is illustrated in Figure 13. The unit dispatch takes place considering the following assumptions:

- CHP plants are operated in must-run mode, which means that they should operate throughout the year to ensure the security of heating supply. The CHP plants with condensing-extraction turbines are considered to generate at least as much electric power as required to supply its assigned set heat output value. In some system variations, some CHP plants also contribute to the coverage of the residual load (e.g. variations 7 to 11).

  The assumption that CHP plants are not switched off at any time is justified by the following argument: according to the mass flow set values of the DHG shown in Figure 12, a certain amount of heating water mass flow should be provided at all times at the east and west sites, even in times with minimum heat demand. Turning off the CHP plants at these sites would mean having to pump cold return water into the supply line. To avoid that, the CHP plants are not switched off in this work at any time, not even in hot summer days. In such days, from a fuel consumption and CO_2 emissions point of view, it does not make any substantial difference to send the live steam directly to the heating condensers (bypassing the steam turbines) or sending the live steam through the turbines while producing minimum power.

- The RE sources have feed-in priority in the electricity grid. Excessive RE production is only curtailed in moments of negative residual load. Residual load is defined here as the difference between the net electricity demand and the RE production. The net electricity demand is in turn defined as the total demand minus the must-run production.
- The positive residual load is covered mainly by the coal-fired power plant at the south site (variations 1 to 9) and the GT power plant at the south site (variations 10 and 11). Plants with lower power production costs (based on fuel costs and electrical efficiency) have always priority. Plants with higher power production costs are activated only if required by the residual load. As said before, in variations 7 to 11, some CHP plants also contribute to the coverage of the residual load.

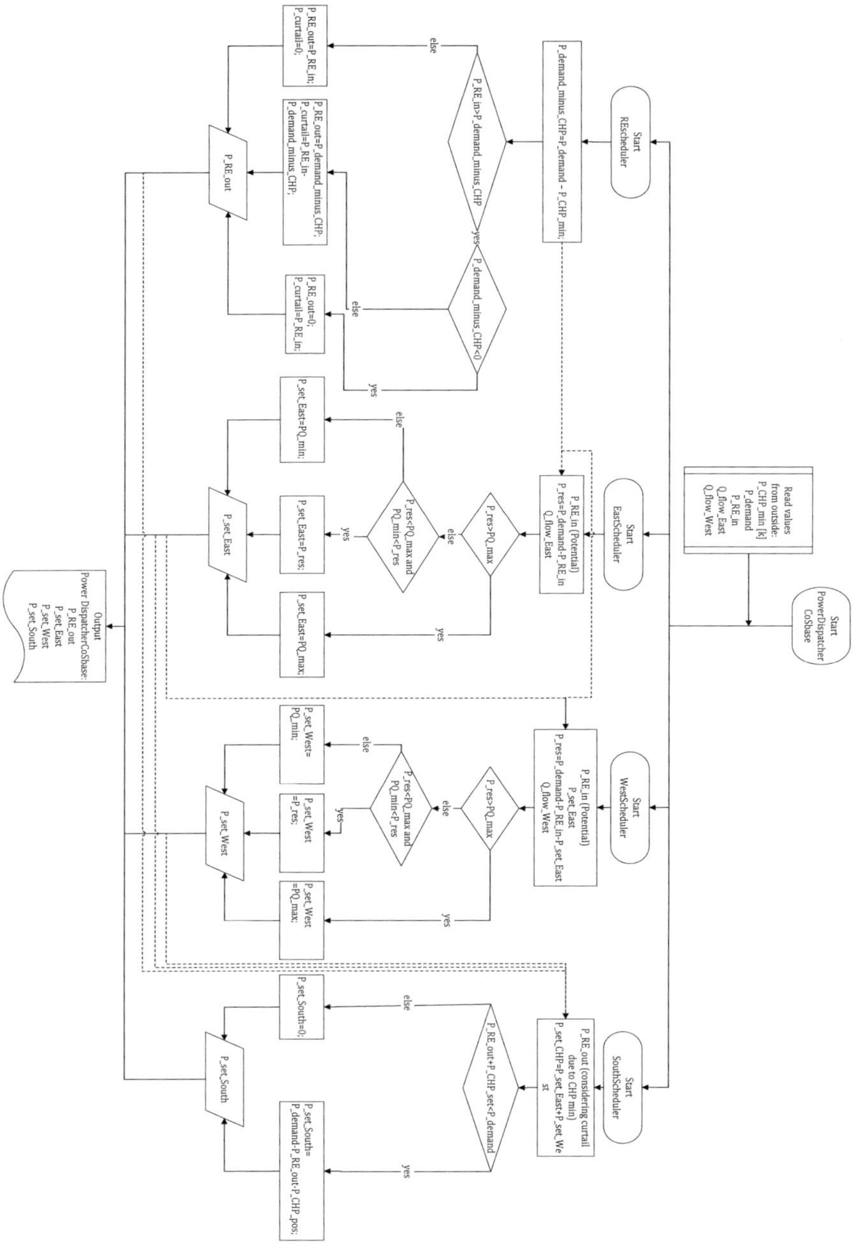

Figure 13 – Power production dispatcher: sample logic structure

4.4 CO_2 emissions, economic and energetic analysis

Besides analyzing the production profiles resulting from the simulation of the different system variations, these can be analyzed and compared with a set of relevant annual performance indicators, such as the total annual CO_2 emissions, total annual costs, or share of RE and CHP power in electricity consumption. This section describes the calculation of these values.

4.4.1 CO_2 emissions analysis

The total annual CO_2 emissions of the system, in tons per year, are calculated by multiplying the fuel-specific CO_2 emission factors (see Table 5) by the fuel consumption of each of the power or heat producers fired with fossil fuels, as expressed in the following equation

$$m_{CO_2,total} = \sum_{i=1}^{n} e_{fuel,i} \cdot Q_{fuel,i} \tag{23}$$

where n is the number of power and heat producers fired with fossil fuels. $e_{fuel,i}$ is the fuel-specific CO_2 emission factor and $Q_{fuel,i}$ is the fuel consumption of the ith producer. The CO_2 emissions incurred for the construction or production of the components is not considered. Since the total annual CO_2 emissions of the system are considered to be the most relevant quantity, the CO_2 emissions are not specifically allocated to the electricity or the heat consumption.

Table 5 – Fuel-specific CO_2 emission factors (e_{fuel})

Fuel type	Fuel-specific CO_2 emission factor [kg/MWh]
Hard coal	337
Natural gas	202
Light fuel oil	266
Waste	162
Biomass, wind onshore, wind offshore, photovoltaic, run-off hydro	0

4.4.2 Economic analysis

The total annual costs, in € million per year, are calculated by adding up the costs of every component of the energy park. These components include all electricity and heat producers required to satisfy the system's energy demand as well as all RE integration technologies. The costs elements considered in the calculation are the fixed costs and the variable costs. Fixed costs are the investment costs' annuities and the fixed operation and maintenance (O&M) costs. Variable costs are the fuel costs, the CO_2 certificate costs and variable O&M costs. The calculation of the total annual costs is described in the following equation

$$\begin{aligned} C_{total} &= \sum_{i=1}^{n} (C_{inv,i} + C_{O\&M\ fix,i}) \\ &+ \sum_{i=1}^{n} (C_{fuel,i} + C_{CO_2 certificates,i} + C_{O\&M\ var,i}) \\ &= \sum_{i=1}^{n} (c_{inv,i} \cdot P_{nom,i} \cdot A_i + c_{O\&M\ fix,i} \cdot P_{nom,i}) \\ &+ \sum_{i=1}^{n} (c_{fuel,i} \cdot Q_{fuel,i} + c_{CO_2\ certificate} \cdot m_{CO_2,i} + c_{O\&M\ var,i} \cdot E_{out,i}) \end{aligned} \tag{24}$$

where n is the number of all system components. $C_{inv,i}$ is the annuity of the fixed investment costs, $C_{O\&M\ fix,i}$ are the fixed operation and maintenance costs, $C_{fuel,i}$ are the variable costs of fuel, $C_{CO_2 certificates,i}$ are the variable costs of CO_2 certificates, and $C_{O\&M\ var,i}$ are the variable O&M costs of the ith system component. $c_{inv,i}$ are the specific fixed investment costs, $c_{O\&M\ fix,i}$ are the specific fixed operation and maintenance costs, P_{nom} is the nominal power or heat output, and A_i is the annuity factor of the ith system component. $c_{fuel,i}$ are the specific fuel costs, $Q_{fuel,i}$ remains as defined in equation (23), $c_{CO_2\ certificate}$ are the specific costs of CO_2 certificates, $m_{CO_2,i}$ are the CO_2 emissions, and $c_{O\&M\ var,i}$ are the specific variable O&M costs of the ith system component.

To allow a better comparability of the system variation, it is assumed that for every variation, the whole energy park has to be built from scratch, i.e. existing plants or retrofits to existing plants are considered to be new plants which require full investment. The different coefficients used for this calculation are presented in Table 6. The values in this table are based on different sources such as [48, 78–84] and on own assumptions.

One particular figure on this table must be justified at this point: the specific "fuel costs" assumed for the electrode boilers and the electrolyzers are considered to be zero, since these units are operated with RE surplus. Negative power prices and other electricity price components such as grid-use fees (*Netznutzungsentgelt*), CHP-fee (*KWK-Umlage*), offshore-fee (*Offshore-Umlage*), DSM-fee (*Abschlaltbare Lasten Umlage*), RE-fee (*EEG-Umlage*) are not considered in this work. In fact, since 2017, the § 13-6a of the EnWG states that 2 GW of PtH capacity are allowed and the costs required for the investment in PtH units will be reimbursed to the CHP operators by the TSO. The TSO will even reimburse the electricity costs incurred by the operators of these so-called switchable loads (*Anschaltbare Lasten*). And according to the aforementioned paragraph, if the 2 GW PtH capacity is not achieved, other technologies would be considered, which suggests that the same conditions (investment support and electricity price exemption) could apply to PtG plants in the future. This would very well justify the specific "fuel costs" of zero assumed for PtG and PtH units.

It is assumed that the waste incineration plant has also specific fuel costs of zero. This kind of plants are usually operated by a municipal or private waste-gathering company, which in fact charges their costumers for the service of picking up and getting rid of their garbage. To avoid using negative costs (i.e. incomes) for this work's calculations, the specific fuel costs of zero are selected.

Even though the PtG, PtH and waste incineration plants have specific fuel costs of zero, the rest of the costs components (e.g. investment costs and O&M costs) are considered in the calculation of the system's total annual costs.

Table 6 – Parameters for costs calculations

	Spec. inv.	Spec. inv.	Spec. inv.	Spec. inv.	Fix O&M	Fuel	CO_2 cert.	Var. O&M	Annuity term
	€/ W	€/ MWh	€/ m^3	€/ m	€/ W	€/ MWh	€/ t-CO2	€/ MWh	years
PV	1.30	-	-	-	0.035	0.0	-	0.0	20
Wind onshore	1.17	-	-	-	0.050	0.0	-	18.0	20
Wind offshore	3.00	-	-	-	0.120	0.0	-	35.0	20
Biomass	2.50	-	-	-	0.005	3.8	-	19.9	20
Run-off hydro	1.60	-	-	-	0.032	0.0	-	0.0	50
Coal-fired power plant	1.50	-	-	-	0.023	10.2	6	1.3	30
Coal-fired CHP plant	1.50	-	-	-	0.023	10.2	6	1.3	30
CCGT-CHP plant	0.65	-	-	-	0.005	25.0	6	3.5	20
Gas turbine plant	0.50	-	-	-	0.003	25.0	6	5.0	20
Pumped hydro	0.73	0.0	-	-	0.003	25.0	-	5.0	20
Gas-fired heater	0.24	-	-	-	0.005	25.0	6	-	20
Waste incineration plant	0.24	-	-	-	0.005	0.0	-	-	30
Electrode boilers	0.07	-	-	-	0.005	0.0	-	-	20
Electrolyzers	0.90	-	-	-	*	0.0	-	-	20
Hydrogen cavern	-	-	60	-	*	-	-	-	30
Hydrogen grid	-	-	-	1220	*	-	-	-	20
District heating grid	-	-	-	3000	*	-	-	-	20
Gas-fired heaters	0.122	-	-	-	*	25	-	-	20
Oil-fired heaters	0.122	-	-	-	*	28.3	-	-	20
Electric heaters	0.474	-	-	-	*	29	-	-	20
District heat	0.995	-	-	-	*	25	-	-	20
Other heaters	0.995	-	-	-	*	25	-	-	20
Interest rate (all)	0.07								
Notes	* Fix O&M costs are assumed to be 5 % of the investment costs								
Sources	Based on [48, 78–84] and on own assumptions								

4.4.3 Energetic analysis

In the energetic analysis, the system variations are compared with the help of three performance indicators: the share of RE in electricity consumption, the share of CHP power in electricity consumption and the share of RE in district heating consumption. These quantities are calculated in every variation and are defined as

$$RE_{share,el} = \frac{E_{el,RE,cons} + E_{el,H_2,out}}{E_{el,cons}}, \tag{25}$$

$$CHP_{share,el} = \frac{E_{el,CHP,out}}{E_{el,cons}}, \tag{26}$$

$$RE_{share,DH} = \frac{Q_{th,RE}}{Q_{th,DH}} = \frac{0.5 \cdot Q_{th,waste,out} + Q_{th,PtH,out} + Q_{th,H_2,out}}{Q_{th,DH,cons}}, \tag{27}$$

where $RE_{share,el}$ is the share of RE in electricity consumption, $E_{el,RE,cons}$ is the part of the RE potential which is directly consumed, $E_{el,H_2,out}$ remains as defined in equation (13), $E_{el,cons}$ remains as defined in equation (15), $CHP_{share,el}$ is the share of CHP power in electricity consumption, $E_{el,CHP,out}$ is the energy produced by co-generation in all CHP plants, $RE_{share,DH}$ is the share of RE in district heating consumption, $Q_{th,RE}$ is the heat produced with RE sources, $Q_{th,DH}$ remains as defined in equation (16), $Q_{th,waste,out}$ is the heat produced by waste incineration plants (50 % of which is considered to be of biogenic nature according to[67]), $Q_{th,PtH,out}$ is the heat produced by PtH units, and $Q_{th,H_2,out}$ remains as defined in equation (14).

For the calculation of the energy produced by co-generation in all CHP plants ($E_{el,CHP,out}$), the calculation method involving the CHP coefficient described in section 4.2.2 is applied. The terms involving the PtH unit and the hydrogen co-firing terms are used only in the applicable system variations. The terms regarding the hydrogen co-firing ($E_{el,H_2,out}$ and $Q_{th,H_2,out}$) were described in section 4.2.3.

Besides the aforementioned performance indicators, the definitions of efficiency terms used in this work are summarized in Table 7. Although these efficiency terms were presented in previous sections, they are summarized here because they are relevant for the model description.

Table 7 – Summary of efficiency terms

Efficiency term	**Defining equation**	**Equation number**
Electrical efficiency	$\eta_{el} = \frac{P_{el,PP,out}}{\dot{Q}_{fuel,PP,in}}$	(1)
Heat efficiency	$\eta_{th} = \frac{\dot{Q}_{th,HP,out}}{\dot{Q}_{fuel,HP,in}}$	(2)
Electrical efficiency of a CHP plant	$\eta_{el,CHP} = \frac{P_{el,CHP,out}}{\dot{Q}_{fuel,CHP,in}}$	(3)
Heat efficiency of a CHP plant	$\eta_{th,CHP} = \frac{\dot{Q}_{th,CHP,out}}{\dot{Q}_{fuel,CHP,in}}$	(4)
Fuel utilization efficiency of a CHP plant	$\xi_{CHP} = \frac{P_{el,CHP,out} + \dot{Q}_{th,CHP,out}}{\dot{Q}_{fuel,CHP,in}}$	(5)
Power to Heat efficiency	$\eta_{th,PtH} = \frac{\dot{Q}_{th,PtH,out}}{P_{el,PtH,in}}$	(6)
Hydrogen production efficiency of the electrolyzer	$\eta_{Electrolyzer} = \frac{\dot{m}_{H_2,ELY,out} \cdot LHV_{H_2}}{P_{el,ELY,in}}$	(7)
PtP efficiency of a PtGtCHP plant	$\eta_{PtP,PtGtCHP} = \frac{P_{el,H_2,out}}{P_{el,ELY,in}}$	(17)
PtH efficiency of a PtGtCHP plant	$\eta_{PtH,PtGtCHP} = \frac{\dot{Q}_{th,H_2,out}}{P_{el,in,Electrolyzer}}$	(18)
PtCHP efficiency of a PtGtCHP plant	$\xi_{PtCHP,PtGtCHP} = \frac{P_{el,H_2} + \dot{Q}_{th,H_2}}{P_{el,in,Electrolyzer}}$	(19)
Electrical efficiency of a PtGtCHP plant	$\eta_{el,PtGtCHP} = \frac{P_{el,CHP,out}}{P_{el,ELY,in} + \dot{Q}_{NG,in}}$	(20)
Thermal efficiency of a PtGtCHP plant	$\eta_{th,PtGtCHP} = \frac{\dot{Q}_{th,CHP,out}}{P_{el,ELY,in} + \dot{Q}_{NG,in}}$	(21)
Energy utilization efficiency	$\xi_{PtGtCHP} = \frac{P_{el,CHP,out} + \dot{Q}_{th,CHP,out}}{P_{el,ELY,in} + \dot{Q}_{NG,in}}$	(22)

5 RESULTS: PROFILE ANALYSIS

In the following sections, the operation strategies implemented in each of the system variations are analyzed based on the production profiles which result from the yearly simulation of the system models. The yearly power and heat demand profiles presented in Figure 2 and Figure 3, as well as the RE production profiles presented in Figure 4 serve as a basis for the simulations. Although these profiles have time resolutions between 15 minutes and one day (see sections 3.2 and 3.3), Dymola's equation solver uses a variable time step to solve the system of equations resulting from the models. The profile analysis of the system variations includes the analysis of the resulting yearly profiles and the analysis of one selected week. The yearly profiles provide an overview of midterm seasonal effects. The profiles of the selected week facilitate the description of the implemented operation strategy and the analysis of short-term intra-day effects. The week shown is the 8th calendar week, which goes from the 20th to the 26th of February in the reference year 2012.

5.1 Reference system

The reference system consists of RE producers (hydro, biomass, wind and photovoltaics), one gas-fired heating plant at the center site, one coal-fired CHP plant at the east and west sites respectively, and one coal-fired power plant at the south site. The installed RE capacity of the reference system represents the capacity of the year 2012 (see Table 3).

The weekly production profile of the reference system is shown in Figure 14 and can be described as follows: regarding the electricity production profiles, coal-fired CHP plants at the east and west sites are regarded as must-run plants and they produce only as much electricity as required to cover the heat demand. These plants operate steadily during this observation period. The RE sources display a fluctuating production profile which does not exceed the demand profile, therefore no RE curtailment takes place. The production profile of the pumped-hydro plant cannot be clearly identified in this figure because it is only used to compensate small differences between the set values and the actual output of the conventional power producers. The coal-fired power plant at the south site covers the electricity deficit when the RE and the CHP plants together cannot fulfill the demand. Regarding the district heating production profiles, the waste-fired heater and the heat converter at the center site show a constant heat production of 100 MW_{th} in total. The coal-fired CHP plants at the east and west sites display small production fluctuations during the here displayed week, which can be traced back to changes in the ambient temperature profile. In the time frame shown here, the gas-fired heater at the center site operates only at a fraction of its capacity and its heat production is used to cover the peaks of heat demand when the previously mentioned production units cannot cover the total demand completely. The gas-fired peak load heaters at the east and west sites are not in operation during the week shown in this figure.

The yearly production profile of the reference system is shown in Figure 15. Regarding the electricity production, it can be observed that the CHP plants generate in winter more than in summer due to the backpressure operation. The RE production is not sufficient to cover the entire net demand at any moment. The coal-fired power plant at the south site covers the residual load and remains active throughout the year. The production of the pumped-hydro plant cannot be visualized in this figure because of its small contribution. Throughout the year, no RE curtailment takes place in the reference system. Regarding the district heating production profiles, the waste incineration plant at the east site and the heat converter station at the center site produce almost constantly throughout the year. The CHP plants at the east and west sites produce more during the cold winter days at the beginning and end of the year. The gas-fired heaters located at the center site, as well as the auxiliary heaters at the east site, are also in operation during these cold periods of the year. In summer the heat production is evidently less than in winter.

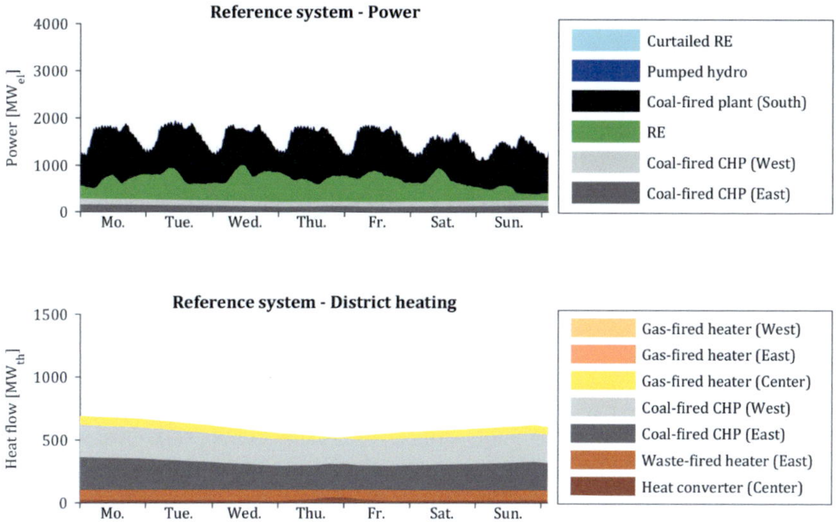

Figure 14 – Reference system: weekly production profile (February 2012)

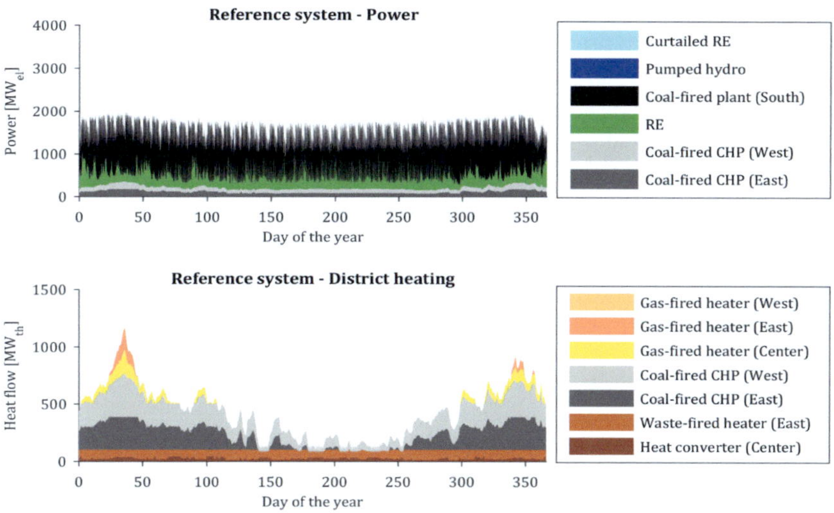

Figure 15 – Reference system: yearly production profile (2012)

5.2 Variation 1: increase in RE production capacity

In the first variation, the energy park is in all aspects equal to the energy park of the reference system, except for the installed RE capacity. The RE production capacity represents that of the year 2050 (see Table 3), and the operation strategy is equal to that of the reference system.

The resulting weekly production profile is shown in Figure 16. Regarding the electricity production, the production profiles of the coal-fired CHP plants remain unchanged with respect to the reference system (see Figure 14). It becomes evident that the RE production increases significantly. At the same time, the power production of the coal-fired power plant at the south site is significantly reduced. However, this plant is still required to cover the residual load in two occasions during the analyzed timeframe. It is also clear that curtailment of RE takes place on a regular basis. The district heating production profiles remain unchanged with respect to the reference system. This means that a) the waste incineration plant at the east site and the heat converter at the center site produce constantly during the analyzed timeframe, b) the coal-fired CHP plants present small fluctuation in their heat production, and c) the gas-fired heater at the center site operates as a peak-load heat producer.

Similar conclusions can be drawn for the yearly production by looking at Figure 17. Regarding the electricity production, the CHP plants also present a strongly seasonal behavior, with peak production in winter and reduced production in summer due to the plants' backpressure operation. The RE production is significantly increased, leading to several periods of RE curtailment throughout the year. It becomes also evident that the RE curtailment is not a particularly seasonal event. Instead, RE curtailment occurs throughout the year without a recognizable pattern. Although the power production of the coal-fired power plant is reduced significantly with respect to the reference scenario (see Figure 15), it is still necessary in periods of time with low RE power output. The power production of the pumped-hydro plant is again so small compared to the total power production, that it cannot be identified in the figure. This is also true for the rest of the variations, which is why this observation will not be repeated in the following sections.

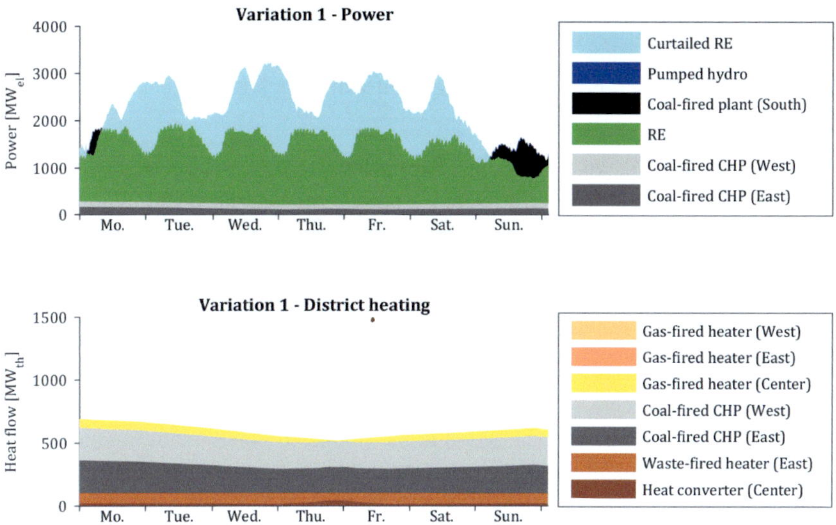

Figure 16 - Variation 1: weekly production profile (February 2050)

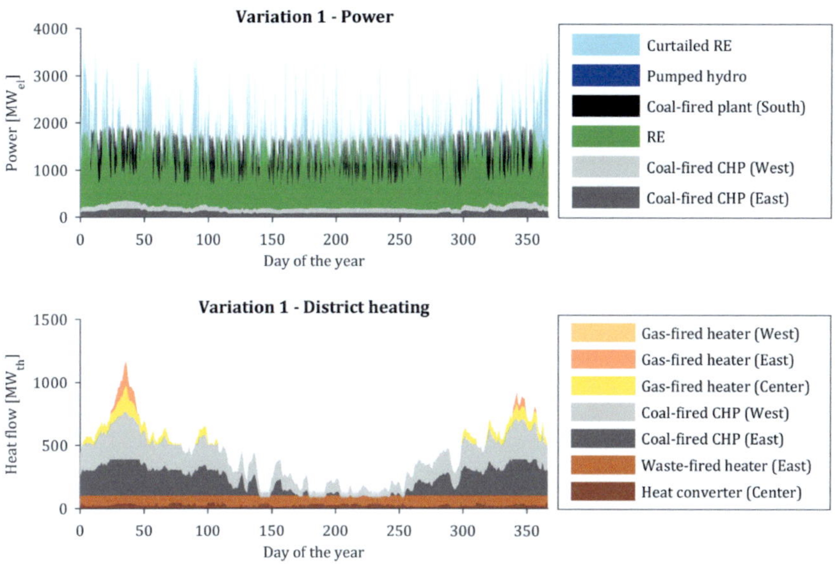

Figure 17 – Variation 1: yearly production profile (2050)

5.3 Variation 2: gas-fired heating plant at the west site

In the system variation 2, the energy park of variation 1 remains unchanged, except for the insertion of a gas-fired heating plant which replaces the coal-fired CHP plant at the west site.

Figure 18 shows the weekly production profiles for this variation. Regarding the electric power production, the production profile of the coal-fired CHP plant at the east site remains basically unchanged. However, the power production of the CHP plant at the west site which was present in the previous variation is now missing. This enables the integration of a larger amount of RE. It also implies a reduction in the RE curtailment, which can be seen by carefully comparing this figure with the weekly profile of variation 1 (Figure 16). This becomes especially evident in days in which the RE curtailment is small, such as Monday, Tuesday or Thursday of the considered week. At the same time, the power production of the coal-fired power plant increases in periods with positive residual load, such as Monday or Sunday of the considered week, because this plant must compensate the missing power output of the CHP plant removed from the west site. Regarding the district heating production profiles, the heat production of the waste incineration plant, the heat converter, the coal-fired CHP plant at the east site, and the gas-fired heater at the center site remain unchanged. However, a gas-fired heater is now in charge of the heat production at the west site, which substitutes the contribution of the removed CHP plant. Although the heat source at the west site is different, the heat production profile of this site remains unchanged.

Figure 19 shows the yearly production profiles of this variation. The same observations done for the weekly profile are valid here. Regarding the electric power profiles, the fraction of the RE production which can be directly used increases due to the removal of the CHP plant at the west site. The RE curtailment decreases slightly and the power production of the coal-fired power plant increases slightly to compensate for the missing CHP plant. Regarding the heat production, it becomes evident that during winter most of the heat production takes place in gas-fired heaters.

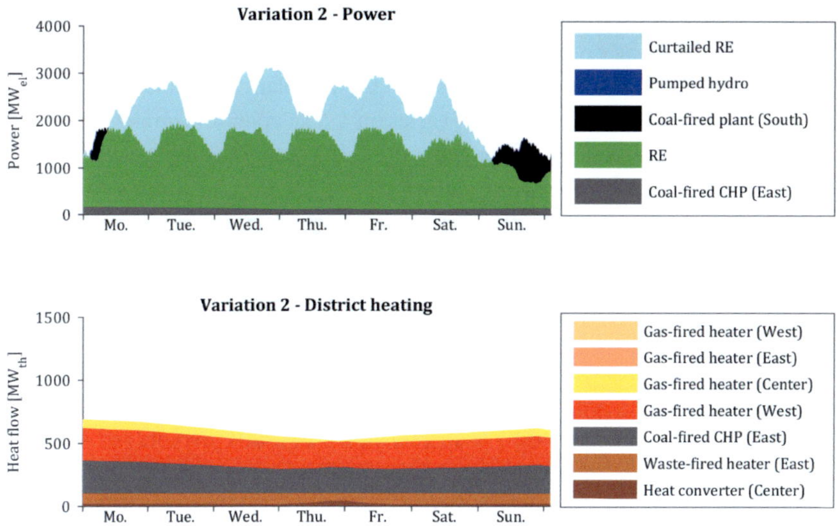

Figure 18 – Variation 2: weekly production profile (February 2050)

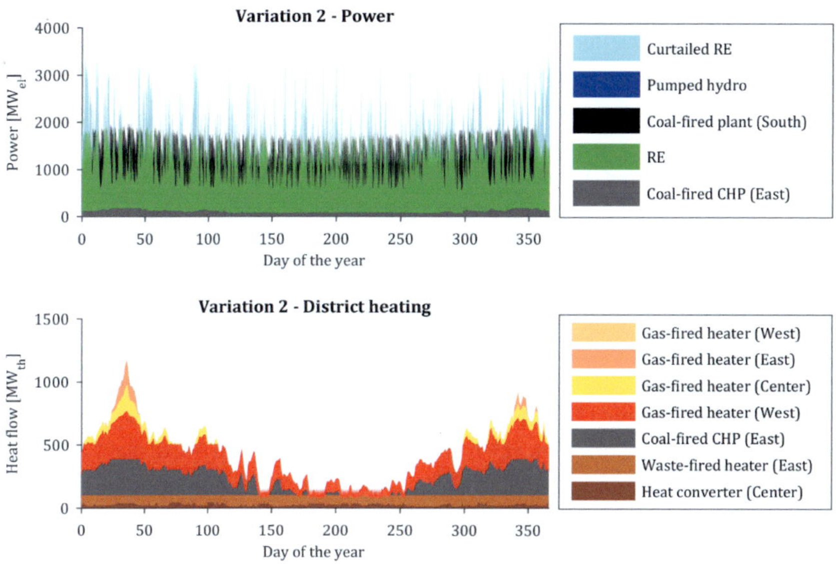

Figure 19 – Variation 2: yearly production profile (2050)

5.4 Variation 3: CCGT-CHP plant at the west site

The weekly and yearly profiles of variation 3 can be found in Figure 20 and Figure 21. The assumption in this variation is that a CCGT-CHP plant is installed at the west site, replacing the coal-fired CHP plant of variation 1. The operation strategy of this plant is the same as the strategy assumed for the CHP plants considered so far: the plant produces only as much power as required to cover the heating demand. This assumption is reasonable considering that the natural gas price is usually higher than the coal price so that burning the expensive fuel should be avoided if possible.

According to stationary simulations, the minimum electric power output of the CCGT-CHP plant is above that of the coal CHP plant of the reference system and variation 1. This has to do with the fact that CCGT-CHP plants have a larger CHP coefficient (in German Stromkennzahl) than their coal-fired counterparts. For this reason, the power production at the west site in variation 3 is higher than in variation 1. This means that, although the share of CHP power in electricity consumption is increased, RE curtailment increases as well, leading to a reduction in the share of RE in electricity consumption. The CCGT-CHP plant provides up to 220 MW_{th} of heat to the west feed-in point. This limit was also obtained through stationary simulations. To maintain the heat output of the west site constant, the capacity of the gas-fired auxiliary heater at this site is increased to 180 MW_{th}.

Figure 20 shows the unchanged production profile of the coal-fired CHP plant at the east site. At the west site, the power production profile of the CCGT-CHP plant is larger than the production of the coal-fired CHP plant being replaced, due to the higher CHP coefficient of the CCGT-CHP plant. This, in turn, leads to an increase in the curtailed RE profile. Regarding the heat production, the profiles remain unchanged, except for a) the heat production at the west site, which now takes place at the CCGT-CHP plant, and b) the auxiliary gas-fired heater at the west site has an output higher than in the previous variations. Similar conclusions can be drawn for the yearly production profiles by analyzing Figure 21. Regarding the power production, the higher power production of the CCGT-CHP plant leads to a reduction in the share of RE which can be used directly, a decrease in the production of the coal-fired power plant and an increase in the curtailed RE power. The seasonality of the heat production profiles is also evident in this variation. The heat production of the CCGT-CHP plant and the gas-fired auxiliary heater at the west site are the only difference between this and the previous variations.

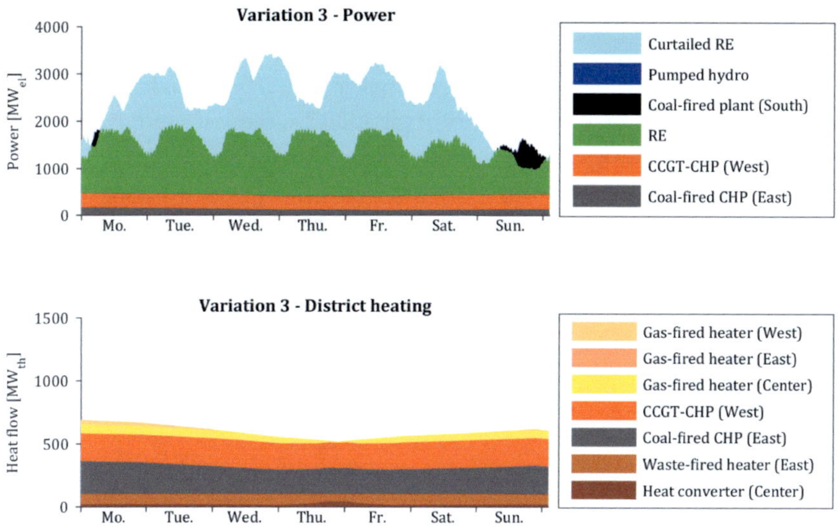

Figure 20 – Variation 3: weekly production profiles (February 2050)

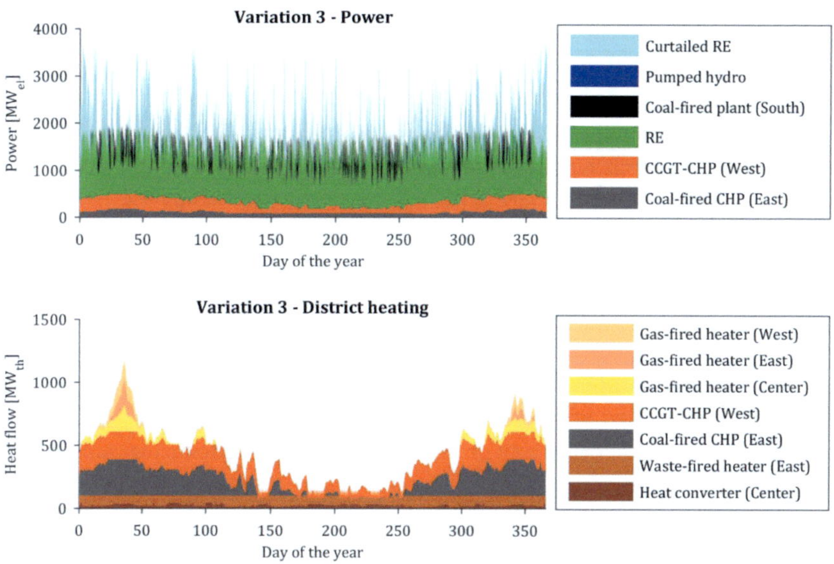

Figure 21 – Variation 3: yearly production profiles (2050)

5.5 Variation 4: PtH unit at the east site

The operation strategy of variation 4 can be explained with the help of Figure 22 and Figure 23. In this variation, an electrode boiler with a heat production capacity of 180 MW_{th} is added to the energy park of variation 3 at the east site. It is assumed that this PtH unit is in series with the heat output of the coal-fired CHP plant. The PtH unit is only operated with otherwise curtailed RE. A PtH efficiency of 1 is assumed for this unit, which means that for every unit of electric power sent to it, one unit of heat is fed into the district heating grid. On the other hand, the operation strategy of the coal-fired CHP plant is altered in such a way that its heat production is reduced when the heat production of the PtH unit is increased.

The heat contribution of the PtH unit and its effect on the heat production of the coal-fired CHP unit at the east site can be clearly identified at the bottom of Figure 22. As seen at the top of this figure, the electricity demand of the PtH unit becomes active when otherwise curtailed RE is present. This, in turn, leads to a reduction in the actually curtailed RE. Additionally, an interesting phenomenon becomes evident by analyzing this figure: during the periods in which the PtH unit is active, the power production of the coal-fired CHP plant is reduced. The reason for this behavior is that the reduced heat production of the CHP plant due to the PtH displacement leads to a reduction in its electricity production. This is true for those periods of time in which the coal-fired CHP plant operates along the backpressure line, which in this case represents the majority of the operation time. This means that the operation strategy described leads to an additional - almost unintentional - increase in the share of RE in electricity consumption. This phenomenon is called the "CHP curtailment effect" in this work. Summarizing, using otherwise curtailed RE power to drive the PtH unit leads to an increase in the share of RE in district heating consumption and, indirectly, to an increase in the share of RE in electricity consumption.

The CHP curtailment effect can also be observed in the yearly production profiles shown in Figure 23, in which the electricity production of the coal-fired CHP plant at the east site is reduced in periods of time in which the PtH unit is active. The PtH unit increases the share of RE in district heating consumption, reduces the amount of curtailed RE power, and decreases the system's CO_2 emissions due to the CHP curtailment effect. The rest of the heat producing and electricity producing components remain unchanged in comparison to variation 3 (compare Figure 23 with Figure 21).

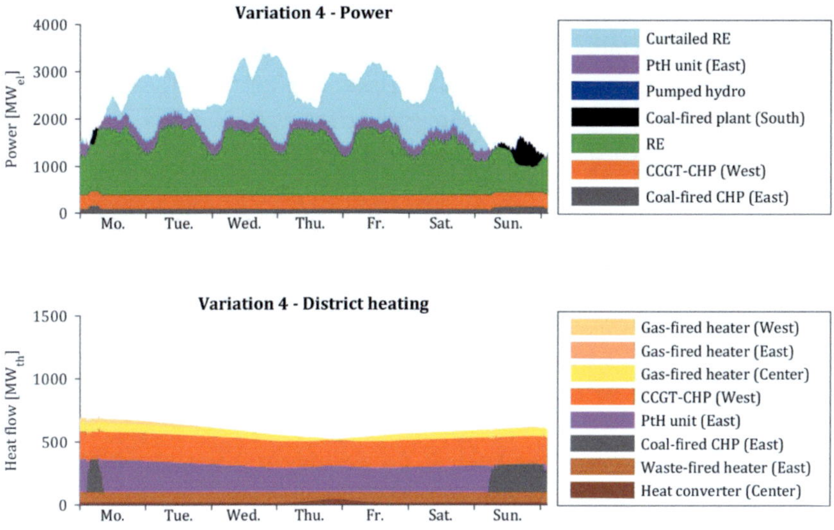

Figure 22 – Variation 4: weekly production profile (February 2050)

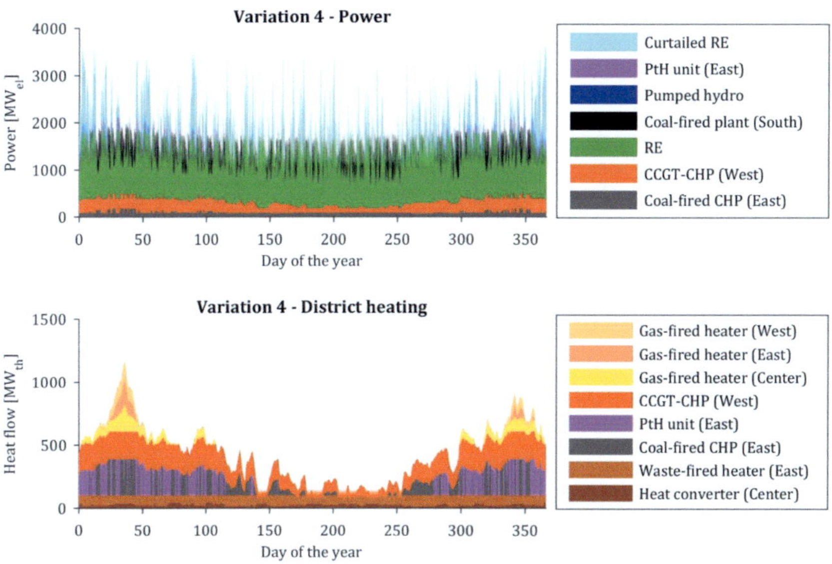

Figure 23 – Variation 4: yearly production profile (2050)

5.6 Variation 5: PtGtCHP plant at the west site

In this variation, additionally to variation 3, a large electrolyzer park and a large underground hydrogen storage are installed close to the west site. The electrolyzer park consists of a cluster of thirty electrolyzer arrays, each having a power rating of 40 MW_{el}, leading to a total nominal power rating of 1,200 MW_{el}. The salt cavern has a nominal volumetric capacity of 500,000 m^3. Of its total hydrogen storage capacity, 37 % is considered to be cushion gas. Regarding the dimensioning of the electrolyzer park, Figure 24 shows the sorted annual load curve of the RE surplus of this variation and the nominal capacity of the electrolyzer park. The RE surplus achieves a maximum value of 2,584 MW_{el}. The nominal capacity of the electrolyzers (1,200 MW_{el}) is selected taking into consideration the available RE surplus. The electrolyzers' nominal capacity lays at 46 % of the maximum RE surplus value. With the current dimensioning, the electrolyzer park operates around 180 days of the year. In order to integrate all the RE surplus in the PtG unit, the electrolyzers' capacity would have to be increased by 54 %-points in size. However, this additional capacity would only be used for around 40 days in the year and would be therefore underutilized. Considering that the investment costs of the electrolyzer park contribute largely to the system's total annual costs, increasing their installed capacity beyond the current level is not advisable.

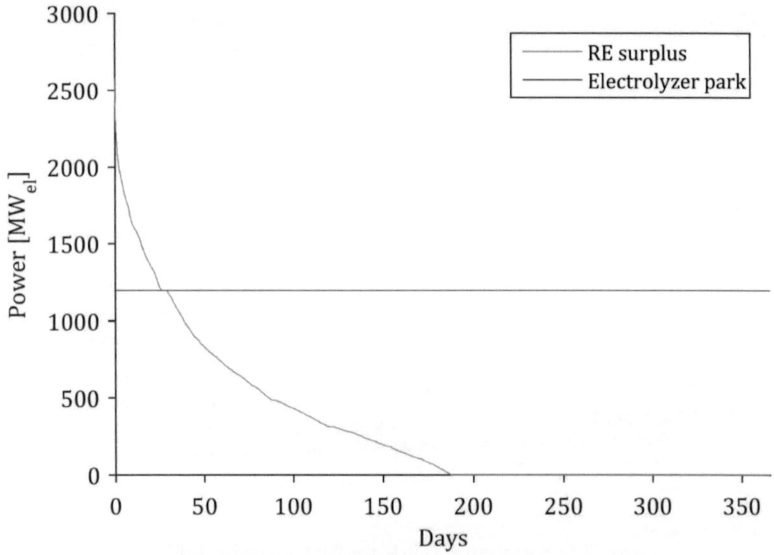

Figure 24 – RE surplus and electrolyzer park installed capacity

The superordinate control of the electrolyzer and cavern system is coupled on the one hand to the RE production dispatcher and on the other hand to the CCGT-CHP plant dispatcher. The electrolyzer produces hydrogen only with otherwise curtailed RE power. Hydrogen is released and sent to the CCGT-CHP plant only if enough working gas is available. The hydrogen sent to the CCGT-CHP plant is then co-fired in the gas turbines, in such a way that 60 % of the heat input from fuel required by the turbines is supplied by the hydrogen combustion. The hydrogen storage was modeled in such a way that the amount of hydrogen at the end of the year is the same as that at the beginning of the year.

The profiles obtained with the described operation strategy are shown in Figure 25 and Figure 26. In these figures, the chemical energy stored in form of hydrogen is also displayed. Only the working gas fraction is shown because the cushion gas fraction must be kept in the cavern at all times. The power production profiles of the coal-fired CHP plant, the CCGT-CHP plant and the RE units are equal to the profiles of variation 3. The coal-fired power plant is still required in periods of time with low RE production. The heat production profiles remain also unchanged. The main difference between this variation and variation 3 is that the PtG plant located at the west site is used to integrate a considerable amount of otherwise curtailed power from RE sources. As defined in section 4.2.3, the RH contribution to power and heat takes place by co-firing hydrogen in the CCGT-CHP plant as long as hydrogen is available. This can be visualized with the help of these figures. Regarding the hydrogen stored in the salt cavern, it becomes evident, that the level of the hydrogen storage increases when the PtG plant is active and the hydrogen production is larger than the hydrogen consumption. In the shown week, hydrogen is consumed at all times. From an annual perspective, the hydrogen storage level increases significantly in periods of time with low heat demand like summer and decreases in periods of high heat demand. The yearly plot suggests that the cavern capacity is well utilized in periods of low heat demand. From around day 160 until day 200 the storage level increases rapidly due to the presence of high amounts of RE production. From around day 200 until day 250 the storage level decreases rapidly because the RE production in this period of time reduces drastically. The hydrogen in the cavern is depleted mainly in periods with high heat demand – when the CCGT-CHP plant produces more heat and power – and low RE potential. Despite these few periods of hydrogen depletion, the PtGtCHP plants effectively contribute to the timely decoupling of RE production and usage.

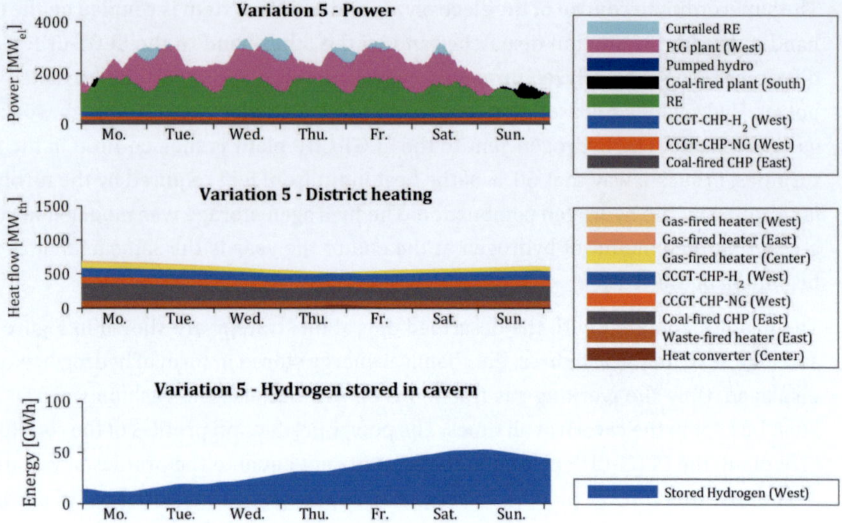

Figure 25 - Variation 5: weekly production profiles (February 2050)

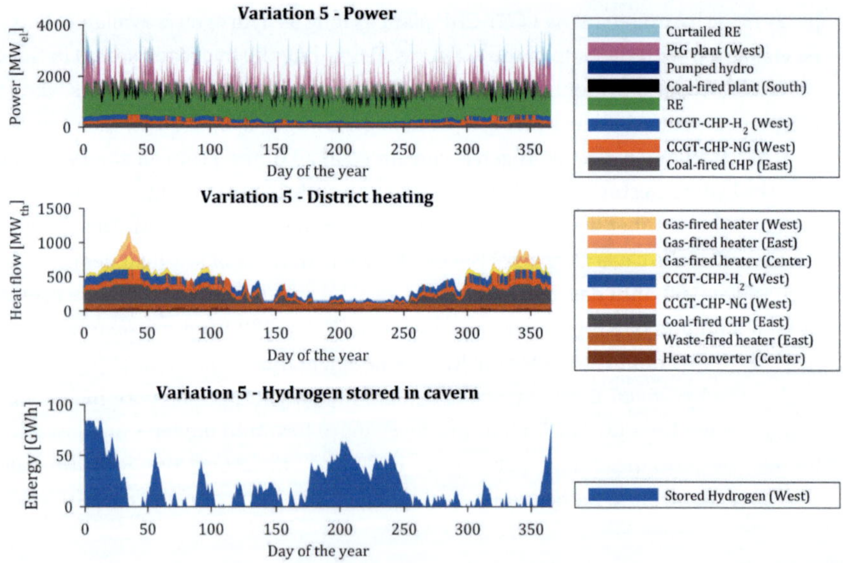

Figure 26 - Variation 5: yearly production profile (2050)

5.7 Variation 6: PtH unit at the east site, PtGtCHP plant at the west site

In this variation, the energy park and operation strategies described in the variations 4 and 5 are merged: a PtH unit is installed at the east site and a PtGtCHP plant is installed at the west site. Both units work with the same "fuel": RE surplus which would be otherwise curtailed. Therefore, an additional merit order is required for these two components. In this work, the PtH unit has priority over the PtG unit. This is justified by the fact that, by using the PtH unit, the consumption of coal is reduced. Once the PtH reaches its maximum capacity, the PtG unit is activated, reducing the natural gas consumption of the CCGT-CHP plant. In this way, the fuel with the higher specific CO_2 emissions, i.e. coal, is substituted first. This is the environmentally favorable solution. However, the economically favorable solution would require the PtG unit to have priority over the PtH unit, since the PtG unit allows a reduction in the consumption of natural gas, which has a higher price than coal in the European fuel price setting assumed in this work.

The results obtained with the described energy park and operation strategy are presented in Figure 27 and Figure 28, which show the power and heat production profiles, as well as the hydrogen storage level. In the weekly plot a combination of the features found in the previous two variations is shown: the PtH unit at the east site operates in the presence of RE surplus and the PtG plant at the west site operates with the remaining RE power potential until reaching their maximum capacities. In the heat plot, the contribution of the PtH unit and the displacement of the coal-fired CHP are shown. The weekly plot of the storage level suggests that the RE power consumed by the PtH unit leads to a reduction in the amount of hydrogen stored in the cavern (compare Figure 27 and Figure 25).

Also in this variation, the coal-fired power plant is still required in periods with low RE production. The RH contribution to power and heat due to hydrogen co-firing in the CCGT-CHP plant can be clearly identified. Despite the presence of the PtH and PtG plant, RE curtailment still takes place in this variation. However, from an annual perspective, the combined action of these units leads to a considerable reduction in the RE curtailment. Besides, seasonal effects similar to the two previous variations can be observed, such as the decrease of the hydrogen storage level in winter and its increase during summer.

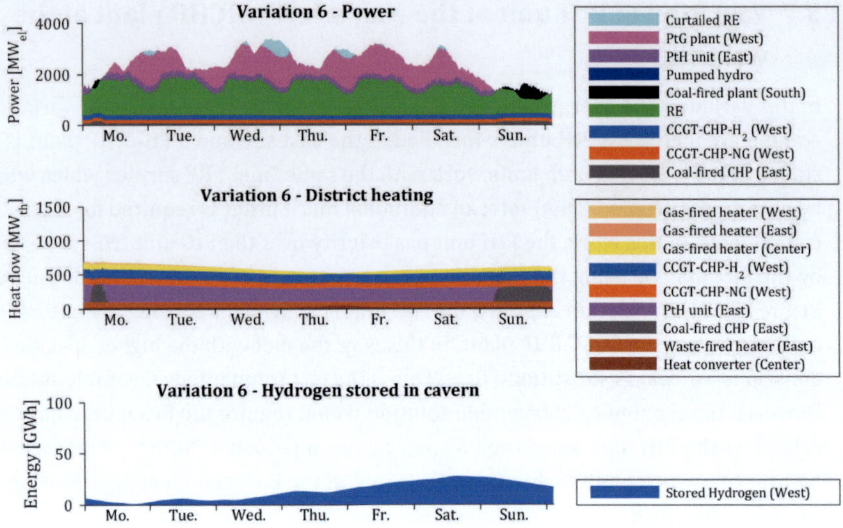

Figure 27 - Variation 6: weekly production profiles (February 2050)

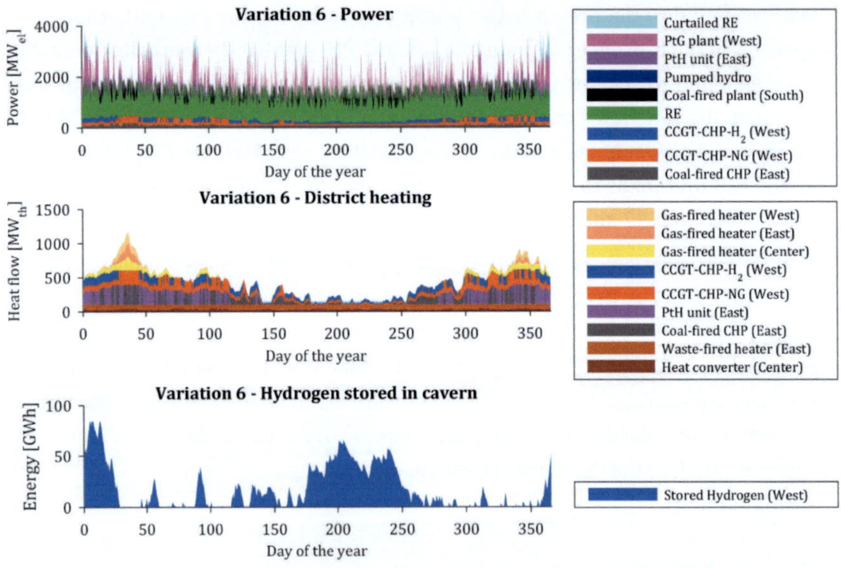

Figure 28 - Variation 6: yearly production profiles (2050)

5.8 Variation 7: CHP plant at the south site replaces plant at the west site

Variation 7, which is the first variation of branch B (see Figure 5), evaluates the effects of converting the coal-fired power plant at the south site into a coal-fired CHP plant which completely substitutes the plant at the west site of the reference system. The results obtained with the selected energy park and operation strategy are shown in Figure 29 and Figure 30. The CHP plant at the south site is modelled as two independent blocks: one block has co-generation capabilities, the other block works as a condensing power station and provides only residual load once the first block reaches its full power production capacity. The coal-fired CHP plant at the east site generates only as much power as required to cover the heating demand (must-run mode). In Figure 29 the coal-fired CHP block at the south site operates steadily to provide its corresponding share of heat. The same can be said of the coal-fired CHP plant at the east site. Compared to the coal-fired CHP plant at the west site being replaced, the CHP block at the south site has a larger must-run profile (compare Figure 29 with Figure 16). This has to do with the fact that the nominal power capacity of the CHP blocks at the west site equals 375 MW_{el}, whereas the nominal power of the CHP block at the south site equals 827 MW_{el}. The minimum-load limits of both plants are defined with respect to the plant's nominal power, which is why the must-run profile of the plant at the south site is higher than that of the plant at the west site. The higher must-run production leads to a reduced share of usable RE (compare also Figure 29 with Figure 16). This, in turn, leads to a higher share of curtailed RE. The residual load is reduced compared to variation 1. Regarding the district heating, the production profiles remain unchanged, except for the fact that the heat contribution of the replaced plant at the west site is covered by the new CHP block at the south site. Similar conclusions can be drawn from analyzing Figure 30. Regarding the electric power production, this variation presents a higher total must-run production, a lower residual load production and a lower share of RE which can be directly used when compared to variation 1. Regarding the district heating production, the production profiles remain almost unchanged when compared to variation 1, except for the fact that, during the cold winter period, the coal-fired CHP plant at the south site achieves a production which is slightly higher than that of the coal-fired CHP blocks being replaced at the west site. This increase in the heat production at the south site leads indirectly to a decrease in the heat production of the CHP plant at the east site in such a way that the total heat production remains constant (compare Figure 30 and Figure 17).

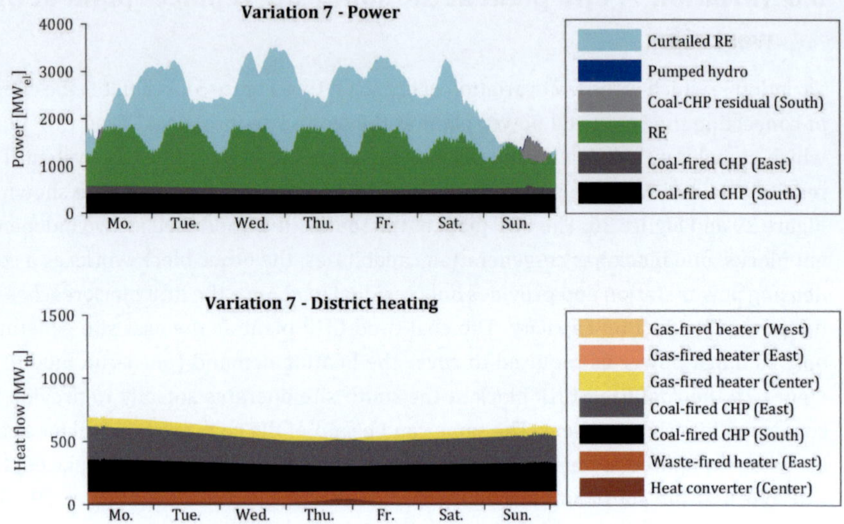

Figure 29 - Variation 7: weekly production profiles (February 2050)

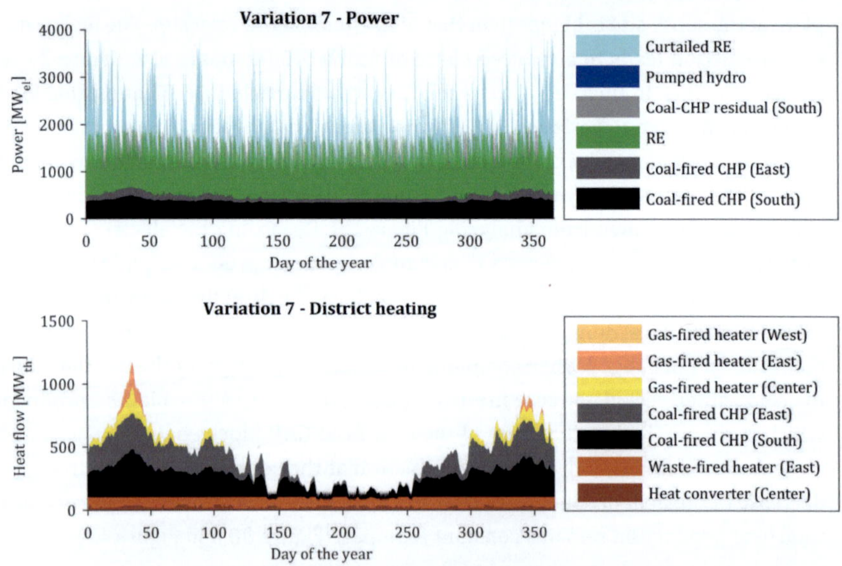

Figure 30 - Variation 7: yearly production profiles (2050)

5.9 Variation 8: CHP plant at the south site, CCGT-CHP plant at the east site

As in the previous variation, in this variation the CHP plant at the south site covers the heat originally supplied by the west site. Additionally, the coal-fired CHP plant at the east site is substituted by a CCGT-CHP plant. The idea behind this is that the total annual CO_2 emissions could be reduced by means of fuel substitution. Retrofit of CHP plants with substitution of fuel from coal to natural gas is financially supported in Germany according to the §7 part 2 of the CHP law [85]. The characteristics of the CCGT-CHP plant of this variation are the same as those of the CCGT-CHP plant in variation 3.

The weekly and yearly production profiles of this variation are shown in Figure 31 and Figure 32. In the weekly profiles it becomes clear that the coal-fired CHP block at the south site produces electricity in the same way as in the previous variation. The CCGT-CHP plant at the east site produces more electricity than its substituted coal-fired counterpart (compare Figure 31 and Figure 29). Consequently, the share of RE which can be directly used is reduced and the amount of curtailed RE is increased. The residual load required is reduced. Regarding the heat production, most of the production profiles remain unchanged, except for the CCGT-CHP production at the east site, which is slightly less than its substituted coal-fired counterpart. This, in turn, leads to an increase in the heat production of the gas-fired heater at the east site, which compensates this difference.

All the above described effects can be found again in the yearly production profile (compare Figure 32 and Figure 30). The coal-fired CHP plant has the same block set-up and operation logic as the previous variation. The CCGT-CHP plant is also able to produce residual power in case the coal-fired CHP block achieves its maximum power production capacity. This can be identified in a few time periods of the yearly power production profiles. The increased contribution of the gas-fired heater at the east site becomes evident in cold winter days, when the total heat production achieves its maximum value.

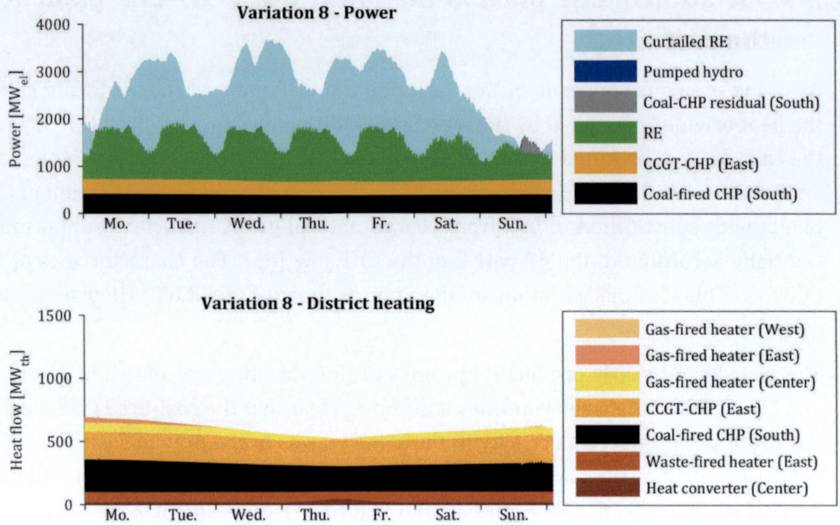

Figure 31 - Variation 8: weekly production profiles (February 2050)

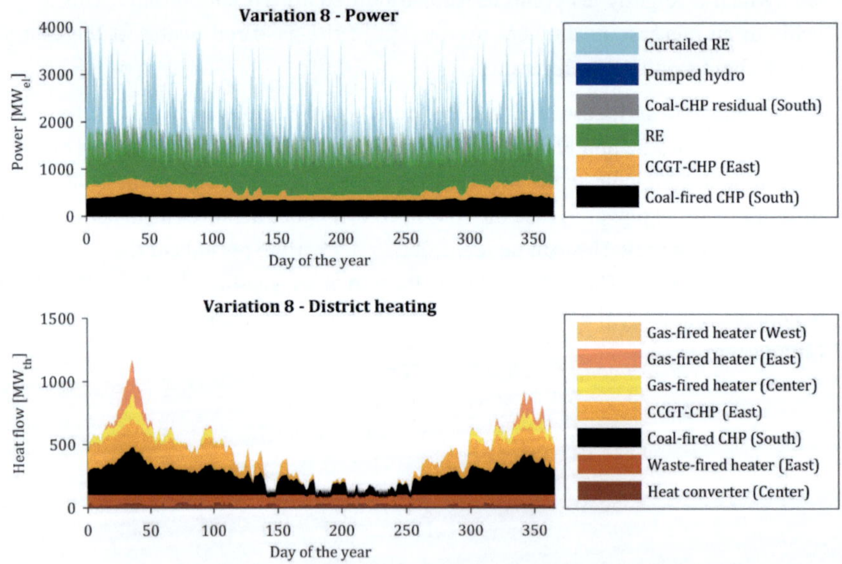

Figure 32 - Variation 8: yearly production profiles (2050)

5.10 Variation 9: CHP and PtH units at the south site, PtGtCHP plant at the east site

The only difference between this variation and variation 8 is that here a PtH unit is installed at the south site and a PtGtCHP plant is installed at the east site. The operation strategy of the PtH and the PtGtCHP units is the same as the one described in variation 6: both units are only operated with otherwise-curtailed RE, the PtH has priority over the PtG unit (displacement of fuel with highest specific CO_2 emissions), the hydrogen is released from the storage only if available and only 60 % of the heat input required by the gas turbines is supplied by combustion of hydrogen.

The weekly and yearly production profiles of this variation are shown in Figure 33 and Figure 34. In the weekly profiles it becomes clear that the switchable demands of the PtH unit and the PtG plant lead to a reduction in the curtailed RE. Besides, the CHP curtailment effect can be identified here again: the displacement of the heat production at the coal-fired CHP block by the PtH unit leads indirectly to a reduction in the power production of the coal-fired CHP plant. Furthermore, the hydrogen level of the cavern storage is clearly dependent on the availability of otherwise curtailed RE and on the fuel consumption of the CCGT-CHP plant, which is in turn mainly determined by the heat demand. The fact that the co-firing of hydrogen in a CCGT-CHP plant leads to an increase in the share of RE in both electricity and district heating consumption can also be seen in this figure.

In the yearly profiles, the CHP curtailment effect mainly takes place in periods with high heat demand. Regarding the hydrogen cavern, the main difference between this variation and variations 5 or 6 is that the cavern is full for a longer period of time in periods of low heat demand. One of the reasons for this phenomenon is that the CCGT-CHP plant is located at the east site, and during summer the waste-fired heating plant provides most of the heat production at this site, leading to reduced hydrogen consumption at the CCGT-CHP plant. Another reason for the higher storage level is that the must-run production of the coal-fired CHP plant at the south site is higher than that of the coal-fired CHP plant in variations 5 and 6 (compare coal-fired CHP contributions in Figure 34 and Figure 26). This leads to an increase in the curtailed RE power and to a higher hydrogen production. It becomes also clear that, under the current assumptions, the storage capacity of the cavern and the availability of otherwise curtailed RE still lead to periods of time in which the stored hydrogen is insufficient for co-firing in the CCGT-CHP plant.

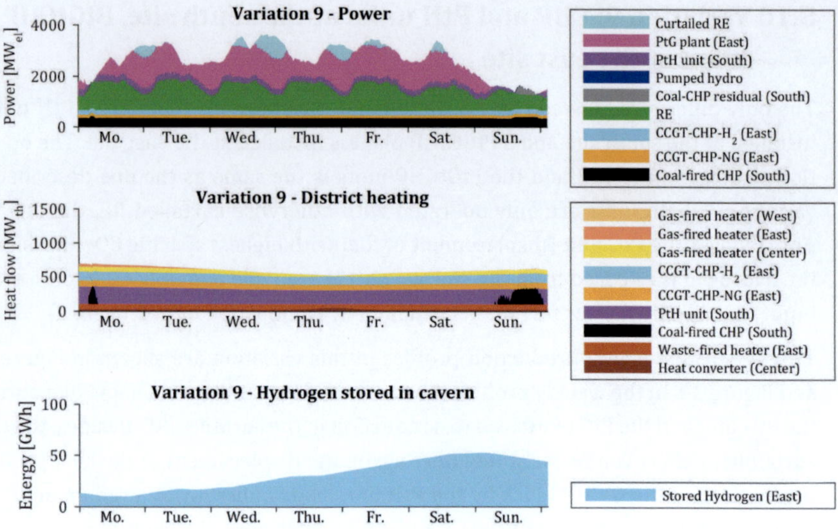

Figure 33 - Variation 9: weekly production profiles (February 2050)

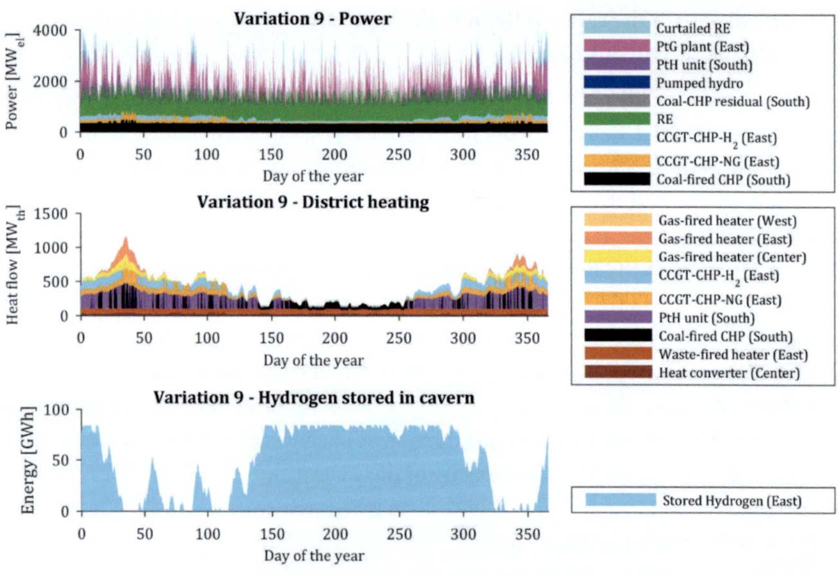

Figure 34 - Variation 9: yearly production profiles (2050)

5.11 Variation 10: GT power plant at the south site, CCGT-CHP plants at the east and west sites

Variation 10, which is the first variation of branch C (see Figure 5) evaluates the consequences of completely removing coal from the analyzed energy system. The coal-fired plant at the south site is replaced by a GT plant consisting of four blocks, each of them with nominal power capacity of 413.5 MW_{el}. CCGT-CHP plants are installed at the east and west sites. The characteristics of the CCGT-CHP plants of this variation are the same as the characteristics of the CCGT-CHP plant in variation 3.

The results obtained with the selected operation strategy are shown in Figure 35 and Figure 36. The CCGT-CHP plant at the east site operates only as a must-run plant, as seen in its power and heat production profiles. The CCGT-CHP plant at the west site is capable of contributing to the residual load coverage, which can also be seen in its weekly and yearly production profiles. Once this plant achieves its maximum production capacity, the GT plant starts to produce power to cover the rest of the residual load, as seen in the weekly power production profile (e.g. on Sunday). If the CCGT-CHP plant at the west site is capable of providing the entire residual load, the GT plant remains inactive (e.g. on early Monday). Due to their higher CHP coefficient, both CHP plants have a higher must-run profile than their coal-fired counterparts, leading to a reduction in the usable RE and an increase in the curtailed RE. In both figures, the increased production of the gas-fired heaters at the east and west sites becomes evident in periods with high total heat production due to the heat production limitations of the CCGT-CHP plants compared to their coal-fired counterparts.

In the yearly production profiles, the contribution of the CCGT-CHP plant at the west site to cover the residual load becomes evident. This is true mainly in periods with large day-night fluctuations of the available RE production. The production profiles of the waste-fired heater at the east site and the heat converter at the center site remain also unchanged in this variation with respect to the previous variations.

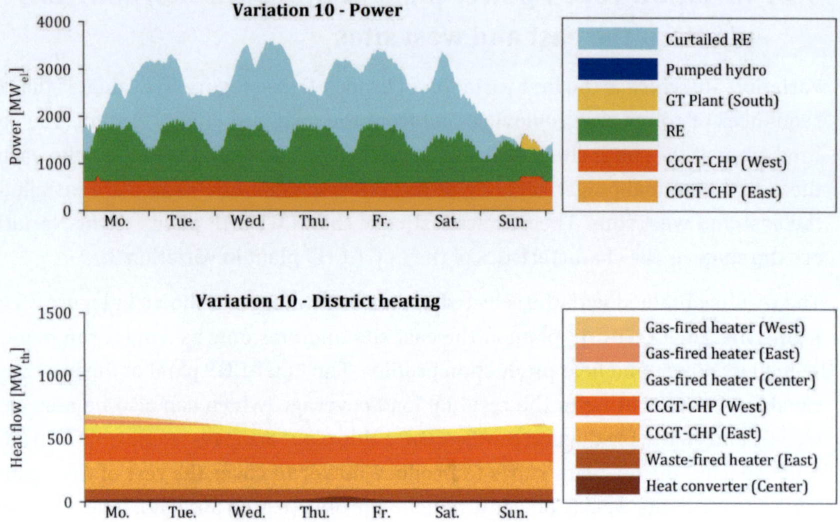

Figure 35 – Variation 10: weekly production profiles (February 2050)

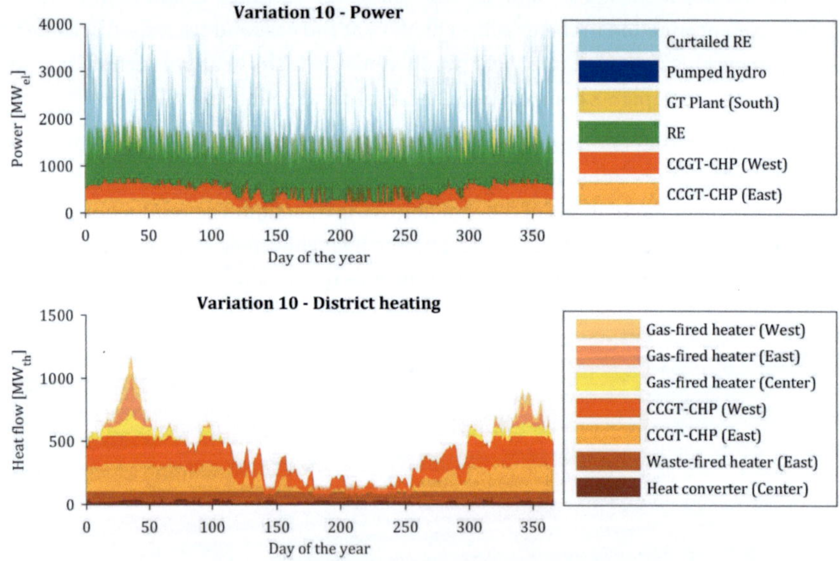

Figure 36 – Variation 10: yearly production profiles (2050)

5.12 Variation 11: GT power plant at the south site, PtGtCHP plants at the east and west sites

This variation has the same energy park as the previous variation: a GT plant at the south site, one CCGT-CHP plant at the east site and one at the west site. However, hydrogen co-firing in the CCGT-CHP plants is assumed in this variation. The hydrogen is produced in two PtG plants and stored in underground salt caverns. Hydrogen production and storage occur close to the east and west sites. Hydrogen is then transported via pipeline to the CCGT-CHP plants (see Figure 6). The total power capacity of the electrolyzes and storage capacity of the salt caverns remains unchanged with respect to the previous variations, but these total capacities are divided equally between both PtG plants: each of them has an electrolyzer power capacity of 600 MW_{el} and a cavern capacity of 250,000 m^3. The results, obtained with the selected energy park and the operation strategy, are shown in Figure 37 and Figure 38. The operation mode of the CCGT-CHP plants is the same as in the previous variation, except for the 60 % hydrogen co-firing scheme already present in other variations with PtGtCHP plants (compare Figure 37 and Figure 35). The RE surplus used for the hydrogen production at the east and west sites is divided equally between both PtG plants. The hydrogen consumed is converted into electricity and heat in both CCGT-CHP plants, as seen in the figures below. The fact that the amount of hydrogen being used is increased and the amount of hydrogen being produced is kept constant leads to a lower utilization of the cavern storage capacities. This is reflected on relatively low storage levels in both the weekly and the yearly plots. Another observable consequence of the above mentioned dimensioning issue is that the hydrogen usage is mainly simultaneous with its production. The yearly profiles put in evidence the behavior of the storage level of both caverns between the seasons: the cavern at the east site achieves higher storage levels in summer because the waste-fired heating plant at this site overtakes most of the heating production in this season, leading to a lower hydrogen utilization by the CCGT-CHP plant. The operation strategy of the PtGtCHP plants leads, in this variation, to an almost simultaneous integration of RE in electricity and heat consumption in periods with RE surplus. To achieve a better temporal decoupling of the hydrogen consumption and production, the operation strategy of the PtGtCHP could be modified. During winter, hydrogen could be only stored in the caverns and not co-fired in the CCGT-CHP plants. In summer, hydrogen could be finally co-fired in the CCGT-CHP plants, leading to a better utilization of the cavern capacity, and to an effective seasonal RE storage solution.

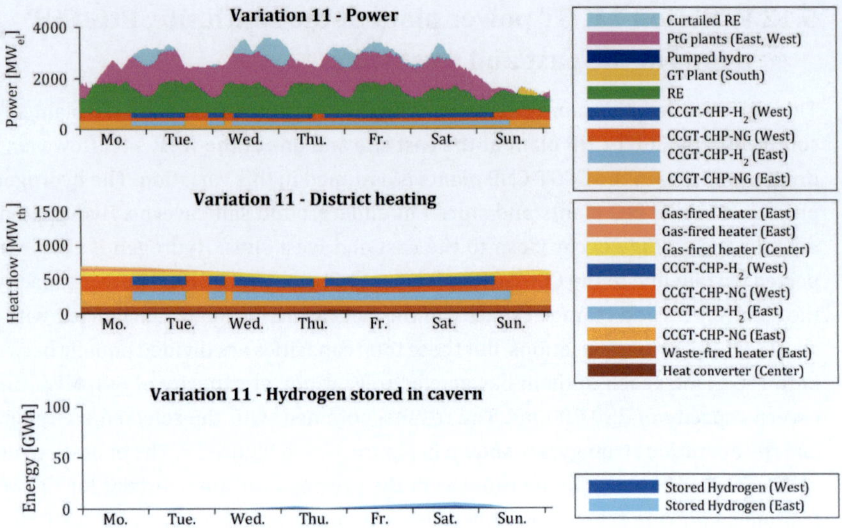

Figure 37 – Variation 11: weekly production profiles (February 2050)

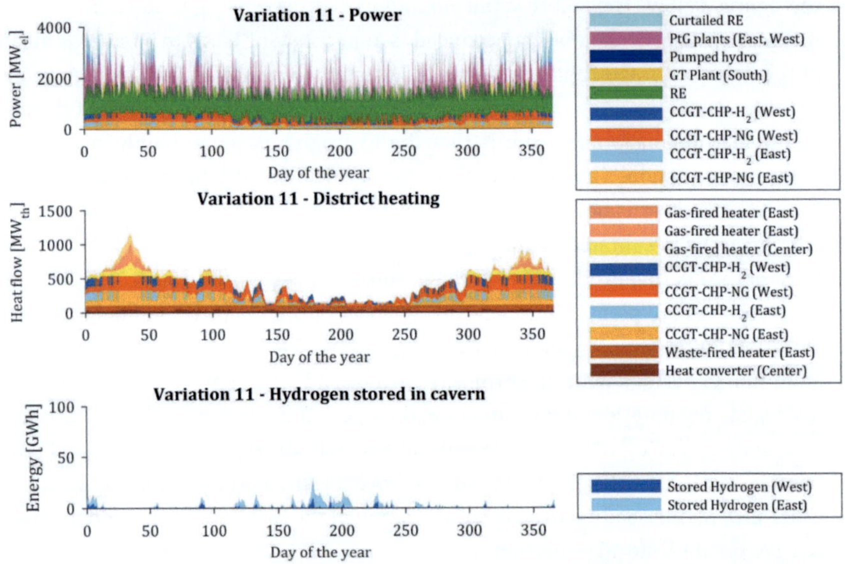

Figure 38- Variation 11: yearly production profiles (2050)

6 RESULTS: ANNUAL ANALYSIS

6.1 Sankey diagrams of representative variations

An overview of the energy and CO_2 emission flows resulting from the annual simulations of the reference system, variation 1, variation 6 and variation 11 is given in Figure 39, Figure 40, Figure 41, and Figure 42 in the form of Sankey diagrams. The variations presented in this section are selected because of the following reasons: a) the reference system and the first variation illustrate the direct effects of increasing the RE production capacity on an otherwise unchanged energy park; b) variation 6 includes both RE integration technologies analyzed in this work, namely PtH and PtGtCHP plants; and c) variation 11 presents low values of total annual CO_2 emissions as well as high shares of RE and CHP power in electricity consumption, which makes it an interesting subject for analysis.

In the Sankey diagrams presented in this section, the energy sources are located on the left and the system's annual electricity and heat consumption, together with the resulting CO_2 emissions, are displayed on the right hand side of the diagrams. The energy conversion units, which convert the different energy sources into heat or electricity, are shown in the middle. As described in section 3.2, the electricity consumption remains constant and equals 12.9 TWh_{el} for all variations. The heat consumption in the reference system is considered to be 22.6 TWh_{th} of which 17 % (3.9 TWh_{th}) is covered by the main district heating grid, 56 % (12.7 TWh_{th}) by gas, 13 % (2.9 TWh_{th}) by heating oil, 5 % (1.1 TWh_{th}) by other heating grids, 8 % (1.8 TWh_{th}) by electric heaters and 1 % (0.2 TWh_{th}) by other sources. This allocation of the heat market shares is based on the values published in [66] and was implemented as such into the model.

According to the Sankey diagram of the reference system, shown in Figure 39, the RE output is entirely used to cover the system's electricity consumption because the RE power output does not exceed the electricity demand at any point in time. As a consequence, no RE is curtailed in this variation. The coal-fired power plant covers the residual load and is the system component with the highest electric energy output, which accounts for 62 % of the electricity requirements of the system. The plant emits 6.1 million tons of CO_2 or 51 % of the total annual CO_2 emissions of this variation. The coal consumption of this plant represents the largest fuel requirement of the system.

The coal-fired CHP plants at the east and west sites have a comparatively small electric energy output because they have a smaller nominal power capacity than the coal-fired power plant and they operate only in must-run mode. Together, these plants require 6.1 TWh of fuel in form of coal and supply 69 % of the main district heating consumption. The waste incineration plant and the central gas-fired heaters are responsible for the rest of the main district heating consumption.

The remaining elements in this figure follow the allocation of the heat market shares described at the beginning of this section. The gas-fired decentral heaters are responsible for 88 % of the gas consumption and 19 % of the CO_2 emissions in this variation. The electric heaters at the top right of the figure are responsible for 8 % of the system's CO_2 emissions, which result from the product of the electric energy input to the heaters and the specific emission factor of the grid.

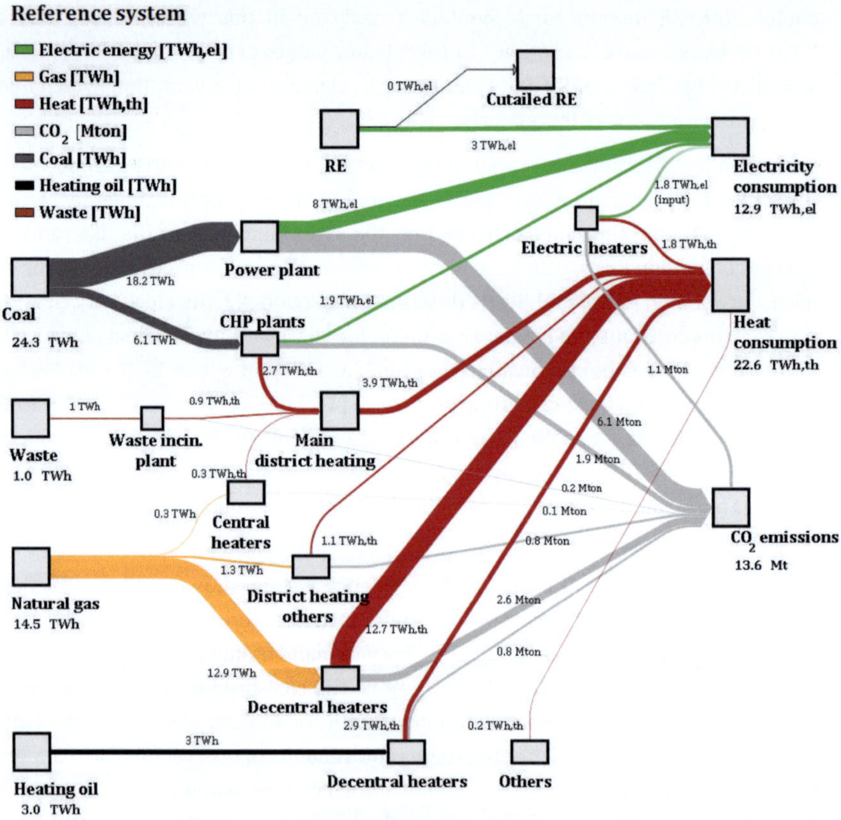

Figure 39 - Sankey diagram of the reference system

According to the Sankey diagram of variation 1, shown in Figure 40, the increased installed RE capacity leads to an increase in the amount of RE directly used to cover the system's electricity consumption from 3 TWh_{el} in the reference system to 9.5 TWh_{el} in variation 1. At the same time, there is an increase in the amount of curtailed RE, which achieves a value of 2.8 TWh_{el}. Another consequence of the increased installed capacity of RE is the reduction in the residual load, which translates into a reduction in the coal-fired power plant's a) electric energy output from 8 TWh_{el} to 1.5 TWh_{el}, b) coal consumption from 18.2 TWh to 3.6 TWh, and c) CO_2 emissions from 6.1 to 1.2 million tons. Due to this reduction of 80 % in the CO_2 emissions, the coal-fired power plant is not the largest CO_2 emitter of the system anymore.

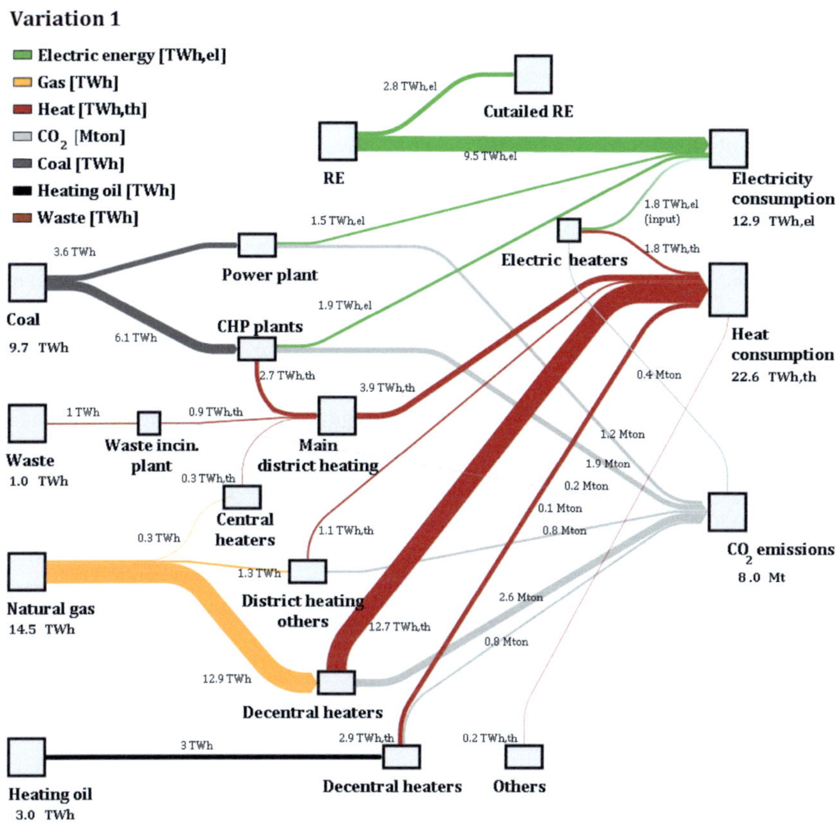

Figure 40 - Sankey diagram of variation 1

The electric energy output, coal consumption and CO_2 emission values of the coal-fired CHP plants do not change in this variation compared with the reference system, because the nominal capacities and the operation strategy of these plants remain unchanged. The energy flows and CO_2 emissions of the main district heating grid and the rest of the system's heat sources remain unchanged except for the electric heaters, which experience a reduction of 64 % in their CO_2 emissions with respect to the reference system because the specific CO_2 emission factor of the electric grid is reduced due to the increased presence of RE. In summary, the total annual CO_2 emissions are reduced in variation 1 mainly due to the displacement of the coal-fired power plant by the RE producers.

Figure 41 shows the annual energy balance of variation 6, in which a PtH unit and a PtGtCHP plant are present. Compared to variation 1, the presence of these units leads to a reduction in the curtailed RE from 2.8 to 0.2 TWh_{el}. The PtH unit receives 0.6 TWh_{el} of RE and converts it into heat. The PtG unit converts 2.6 TWh_{el} of RE into 1.8 TWh in form of hydrogen. The amount of RE directly used to cover the electricity consumption is reduced in this variation with respect to variation 1 by 0.6 TWh_{el}, to a value of 8.9 TWh_{el}, because the coal-fired CHP plant at the west site is substituted by a CCGT-CHP plant, which has a higher CHP coefficient. The hydrogen storage cavern was modeled in such a way that the amount of hydrogen at the beginning and the end of the year was the same, which means that the 1.8 TWh of hydrogen produced in the electrolyzers throughout the year are sent to the CCGT-CHP unit. Besides the CO_2 emission reduction due to fuel substitution from coal to natural gas at the CCGT-CHP plant, the use of RH leads to an additional reduction in the plant's CO_2 emissions, which attain a value of 0.5 million tons, leading to a fuel-specific CO_2 emission factor for the combined fuel over the year of 111 kg/MWh (0.5 million tons divided by 4.5 TWh, which is the sum of the natural gas and hydrogen inputs to the CCGT plant). As a comparison, natural gas alone has a fuel-specific CO_2 emission factor of 202 kg/MWh, as shown in Table 5. The here calculated fuel-specific CO_2 emission factor for the combined natural gas and hydrogen fuel varies from variation to variation, depending on the availability of RE surplus, so that variations with low RE surplus would have higher values and vice versa.

The energy output, heat output, coal consumption and CO_2 emissions of the coal-fired CHP plant decrease with respect to variation 1 because only the CHP plant at the east site is considered and due to the CHP curtailment effect caused by the PtH unit at this site.

The natural gas consumption is increased from 14.6 TWh in variation 1 to 17.4 TWh in variation 6, an increase of 19 %. The reasons for this are the presence of the CCGT-CHP plant and an increase in the fuel consumption of the central heaters, which participate more in the peak-load coverage at the west site in comparison to variation 1 (compare variation 6 with 1 in Figure 5).

The energy flows and CO_2 emissions of the rest of the system's heat sources remain unchanged, except for the electric heaters, which experience a reduction of 73 % in their CO_2 emissions with respect to the reference system, because the specific CO_2 emission factor of the electric grid is reduced due to the fuel substitution, the increased RE output and the RH power contribution.

Summarizing, the CO_2 emission reductions obtained in this variation are mainly due to the fuel substitution (CCGT-CHP plant instead of coal-fired CHP plant), the CHP curtailment effect caused by the PtH unit and the lower fuel-specific CO_2 emission factor of the fuel mix (hydrogen and natural gas) used by the CCGT-CHP plant.

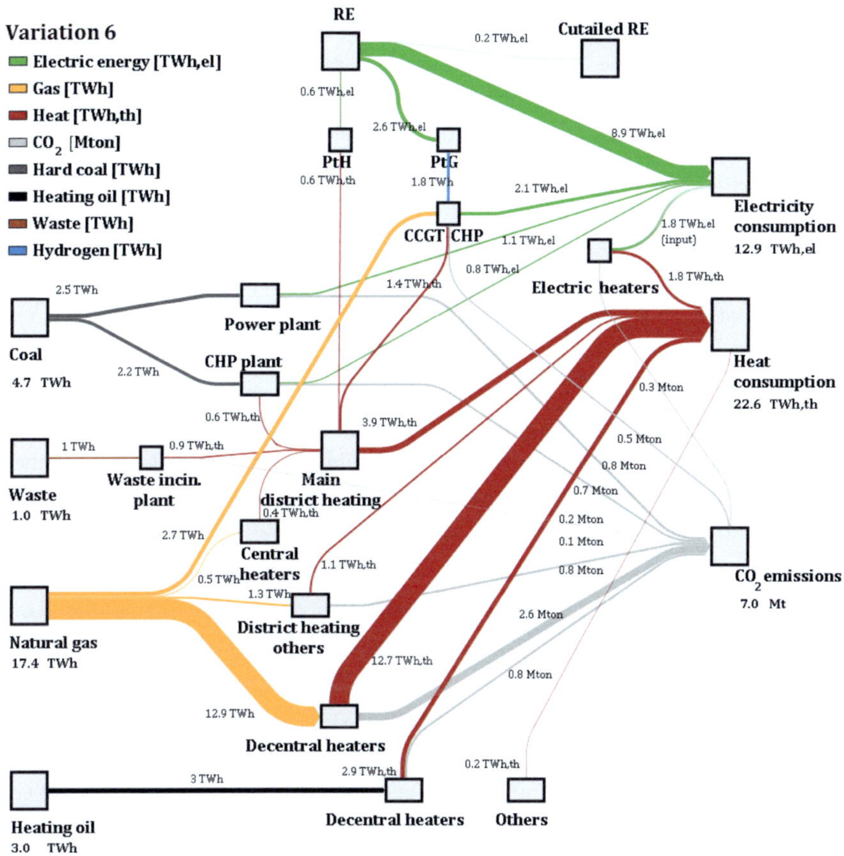

Figure 41 – Sankey diagram of variation 6

Figure 42 shows the Sankey diagram of variation 11. The coal consumption disappears completely in this variation because the coal-fired power plant is substituted by four gas-fired GT blocks, and the coal-fired CHP plants at the east and west sites are substituted by CCGT-CHP plants. As previously mentioned, since the CCGT-CHP plants have a larger CHP coefficient than their coal-fired counterparts, the must-run production of the CCGT-CHP plants is higher. In comparison to variation 6, this leads to a) a decrease of the RE directly used to cover the electricity consumption (here 8.2 TWh_{el}), b) an increase in the curtailed RE (0.4 TWh_{el}), and c) a reduction in the residual load, which is mainly covered by the GT power plants (0.3 TWh_{el}). It becomes clear that the GT power plants are largely oversized, achieving a capacity factor of merely 2.1 %, calculated as the ratio between the plant's annual energy production and the plant's theoretical annual energy production at full load. Decreasing the installed capacity of the GT power plants to 827 MW_{el} would still be a viable solution, but this was not done in this work to enable a better comparability of the different system variations. Reducing the installed GT capacity would lead mainly to a reduction in the total annual costs due to the reduction in the investment costs.

The natural gas consumption increases from 17.4 TWh in variation 6 to 20.4 TWh in variation 11, which represents an increase of 17.2 %. In this variation, 2.6 TWh of hydrogen produced with RE in the PtG plans are co-fired in the CCGT-CHP plants, assuming that the hydrogen cavern storage has the same level at the beginning and the end of the yearly simulation. Without the hydrogen co-firing, the natural gas consumption would be larger.

The energy flows and CO_2 emissions of the rest of the system's heat sources remain unchanged, except for the electric heaters, which experience a reduction of 82 % in their CO_2 emissions with respect to the reference system, because the specific CO_2 emission factor of the electric grid is reduced due to the fuel substitution, the increased presence of RE, and the RH power contribution.

The combined effects of the substitution of coal with natural gas and the presence of PtGtCHP plants lead to a reduction in the annual total CO_2 emissions compared to variation 6 from 7.0 to 6.4 million tons per year, which represents a reduction of 8.6 %. This is in fact the variation with the lowest CO_2 emissions of all.

A more detailed analysis of the conflict between RE and CHP power can be made with the help of the here presented Sankey diagrams. By analyzing the RE flows of variations 1, 6 and 11, the following becomes evident: the RE directly used for the electricity consumption decreases from 9.5 TWh_{el} in variation 1 to 8.9 TWh_{el} in variation 6 and 8.2 TWh_{el} in variation 11. The amount of RE being curtailed in variation 1 equals 2.8 TWh_{el}. Although the amount of RE being curtailed decreases to 0.2 TWh_{el} and 0.4 TWh_{el} in variations 6 and 11 respectively, the amount of RE being used in the RE integration units (PtH and PtG) is actually higher than the 2.8 TWh_{el} curtailed in variation 1. This phenomenon can be explained by the fact that the increase in CHP electrical capacity due to the shift from coal-fired CHP plants to gas-fired CCGT-CHP plants compromises the direct RE utilization. Although the RE integration technologies allow an increase in the share of RE in electricity and district heating consumption, the source of the electricity surplus could be attributed to the increased electric energy output of CHP plants. This increases from 1.9 TWh_{el} in variation 1 to 2.9 TWh_{el} in variation 6 and 4.4 TWh_{el} in variation 11.

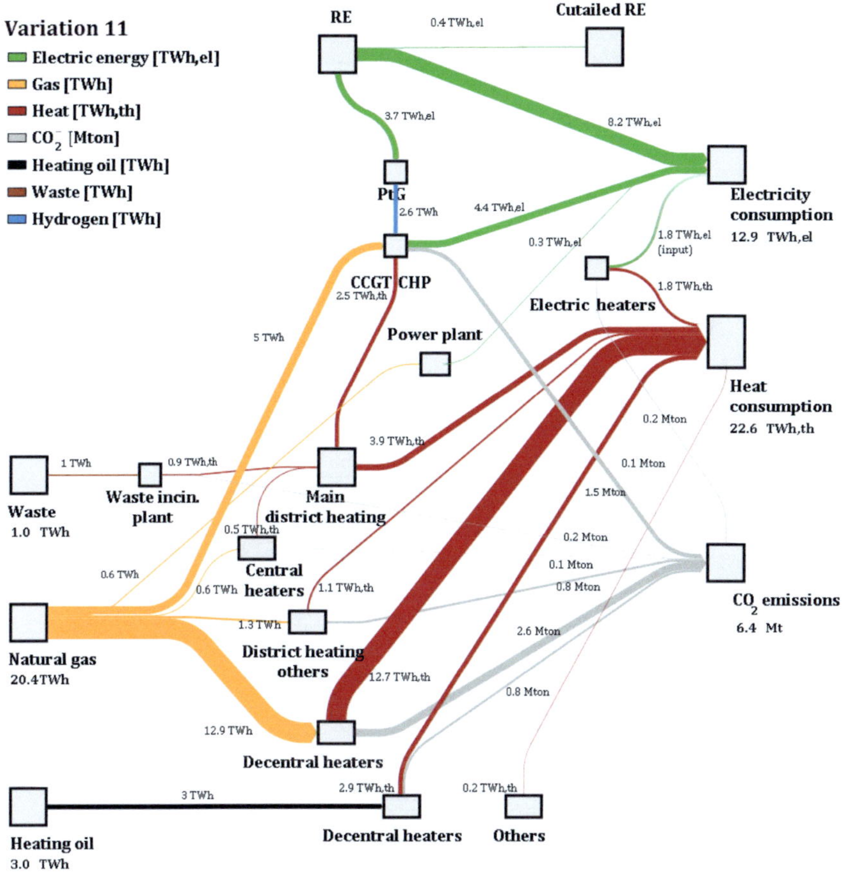

Figure 42 - Sankey diagram of variation 11

6.2 Annual efficiencies of PtGtCHP plants

The partial IO and total IO efficiency sets defined in section 4.2.3 (equations (17) to (22)) can be used to analyze the annual performance of PtGtCHP plants by integrating over one year the terms which define the sets. Figure 43 shows the Sankey diagram of the resulting annual energy inflows and outflows to the PtGtCHP plant in variation 6. These values are extracted from the Sankey diagram shown in Figure 41 and are now used to illustrate the calculation of the annual efficiency sets.

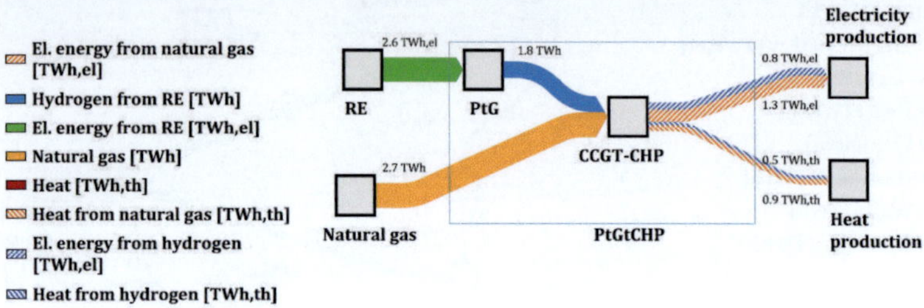

Figure 43 – Sankey diagram of the PtGtCHP plant in variation 6

According to the values in the Sankey diagram and the proposed sets of equations, the annual partial IO efficiency set of these units is calculated as

$$\eta_{PtP,PtGtCHP,a} = \frac{\int_{t=0}^{T} P_{el,H_2,out}\,dt}{\int_{t=0}^{T} P_{el,ELY,in}\,dt} = \frac{E_{el,H_2,out}}{E_{el,ELY,in}} = \frac{0.8}{2.6} = 31\ \% \tag{28}$$

$$\eta_{PtH,PtGtCHP,a} = \frac{\int_{t=0}^{T} \dot{Q}_{th,H_2,out}\,dt}{\int_{t=0}^{T} P_{el,ELY,in}dt} = \frac{Q_{th,H_2,out}}{E_{el,ELY,in}} = \frac{0.5}{2.6} = 19\ \% \tag{29}$$

$$\xi_{PtCHP,PtGtCHP,a} = \frac{\int_{t=0}^{T}\left(P_{el,H_2,out} + \dot{Q}_{th,H_2,out}\right)dt}{\int_{t=0}^{T} P_{el,ELY,in}\,dt} = \frac{E_{el,H_2,out} + Q_{th,H_2,out}}{E_{el,ELY,in}} = \frac{0.8 + 0.5}{2.6} = 50\% \tag{30}$$

where $\eta_{PtP,PtGtCHP,a}$ is the annual PtP efficiency, $\eta_{PtH,PtGtCHP,a}$ is the annual PtH efficiency, and $\xi_{PtCHP,PtGtCHP,a}$ is the PtCHP efficiency of the PtGtCHP plant. $E_{el,H_2,out}$ and $Q_{th,H_2,out}$ are the electricity and heat outputs from hydrogen from of the PtG units as defined in equations (13) and (14), and $E_{el,ELY,in}$ is the electric energy used by the electrolyzer to produce hydrogen.

The annual PtCHP efficiency of the PtGtCHP plant is 50 %, which means that 50 % of the RE sent to the PtGtCHP plant is converted into heat and power. If the same amount of RE would be used to produce RH in a hypothetical electrolyzer park with nominal efficiency of 70 %, related to LHV, and then fired in a hypothetical CCGT power plant with electrical efficiency of 60 % without co-generation, the PtP efficiency would be of 42 % (100 % RE × 70 % electrolyzer efficiency → 70 % energy in form of hydrogen × 60 % CCGT electrical efficiency→ 42 % PtP efficiency) and the PtH efficiency would be of 0 %, since no heat is produced in a CCGT power plant without co-generation. A similar exercise can be done with a hydrogen-fired central heater with heat efficiency of 95 %, in which case the resulting PtH efficiency would be of 67 % (100% RE × 70 % electrolyzer efficiency→70 % energy in form of hydrogen × 95 % heat efficiency of the central heater→67 % PtH efficiency), but the PtP efficiency would be of 0 %, since no power is produced in a central heater. Furthermore, the annual PtH efficiency of the PtGtCHP plant indicates that only 19 % of the RE electricity sent to the PtGtCHP plant is converted into heat. This seems rather small when compared to the PtH efficiency of the PtH unit, which has a value of 100 %. However, the PtGtCHP plant is not only able to make low temperature heat as the PtH unit does, but it is also able to produce electrical power – a better product from an exergetic point of view – with a PtP efficiency of 31 %. Summarizing, PtGtCHP plants enable the usage of RH for the production of electricity or heat with a considerably high annual PtCHP efficiency value.

On the other hand, the annual total IO efficiencies, which take into consideration the total energy inputs and outputs, are also calculated with the values presented in Figure 43, leading to the following results

$$\begin{aligned}\eta_{el,PtGtCHP,a} &= \frac{\int_{t=0}^{T} P_{el,CHP,out}\,dt}{\int_{t=0}^{T} (P_{el,Electrolyzer,in} + \dot{Q}_{NG,in})\,dt} \\ &= \frac{E_{el,CHP,out}}{E_{el,Electrolyzer,in} + Q_{NG,in}} = \frac{2.1}{2.6 + 2.7} = 40\ \%\end{aligned} \tag{31}$$

$$\begin{aligned}\eta_{th,PtGtCHP,a} &= \frac{\int_{t=0}^{T} \dot{Q}_{CHP,out}\,dt}{\int_{t=0}^{T} (P_{el,Electrolyzer,in} + \dot{Q}_{NG,in})\,dt} \\ &= \frac{Q_{CHP,out}}{E_{el,Electrolyzer,in} + Q_{NG,in}} = \frac{1.4}{2.6 + 2.7} = 26\ \%\end{aligned} \tag{32}$$

$$\xi_{PtGtCHP,a} = \frac{\int_{t=0}^{T} (P_{el,CHP,out} + \dot{Q}_{CHP,out})\, dt}{\int_{t=0}^{T} (P_{el,Electrolyzer,out} + \dot{Q}_{NG,in})\, dt} = \frac{E_{CHP,out} + Q_{CHP,out}}{E_{el,Electrolyzer,in} + Q_{NG,in}} = \frac{2.1 + 1.4}{2.6 + 2.7} = 66\ \% \tag{33}$$

where $\eta_{el,PtGtCHP,a}$ is the annual electrical efficiency, $\eta_{th,PtGtCHP,a}$ is the annual thermal efficiency and $\xi_{PtGtCHP,a}$ is the annual energy utilization efficiency of a PtGtCHP plant. $E_{el,CHP,out}$ and $Q_{CHP,out}$ are the electricity and heat outputs of the CHP unit, obtained by integrating the time dependent outputs described in equation (5). $E_{el,Electrolyzer,in}$ remains as described in equation (30), and $Q_{NG,in}$ is the energy in form of natural gas required by the CHP unit to produce electricity and heat.

When considering all energy inputs required for the PtGtCHP plants to operate, the annual electricity efficiency of the PtGtCHP plant reaches 40 %, the heat efficiency reaches 26 % and the unit's total IO fuel utilization efficiency reaches 66 %. These values consider the fact that natural gas can be directly used in the CHP plant, but hydrogen must be produced out of RE via an electrolyzer park, which comes with efficiency penalties.

6.3 Summary of annual results

To enable a comparison between the simulation results of the different variations, a group of relevant quantities was selected. These quantities are: the share of RE in electricity consumption (RE share, el), the share of CHP power in electricity consumption (CHP share, el), total annual costs (Cost (total)), total annual CO_2 emissions (CO_2 (total)), and the share of RE in district heating consumption (RE share, DH). These quantities are defined in detail in section 4.4. The results obtained with the annual simulations are summarized in Figure 44, which displays the system variations grouped into the three main branches defined in section 3.4. These results are analyzed and compared in detail in sections 6.4 and 6.5.

The share of RE in electricity consumption (RE share, el) represents only the share of RE which can be used in the electric grid. This implies that curtailed RE is not considered in this quantity. In variations 5 and 9, where hydrogen from the PtG unit is co-fired in the CCGT-CHP plant, the fraction of the electricity generated by the co-firing of hydrogen is also considered in this term. See equation (25) for details on the calculation of this value.

In this work, the share of CHP power in electricity consumption (CHP share, el) is not the total electricity production of the CHP units. Only the electricity produced in co-generation is considered. See equation (26) for details on the calculation of this value.

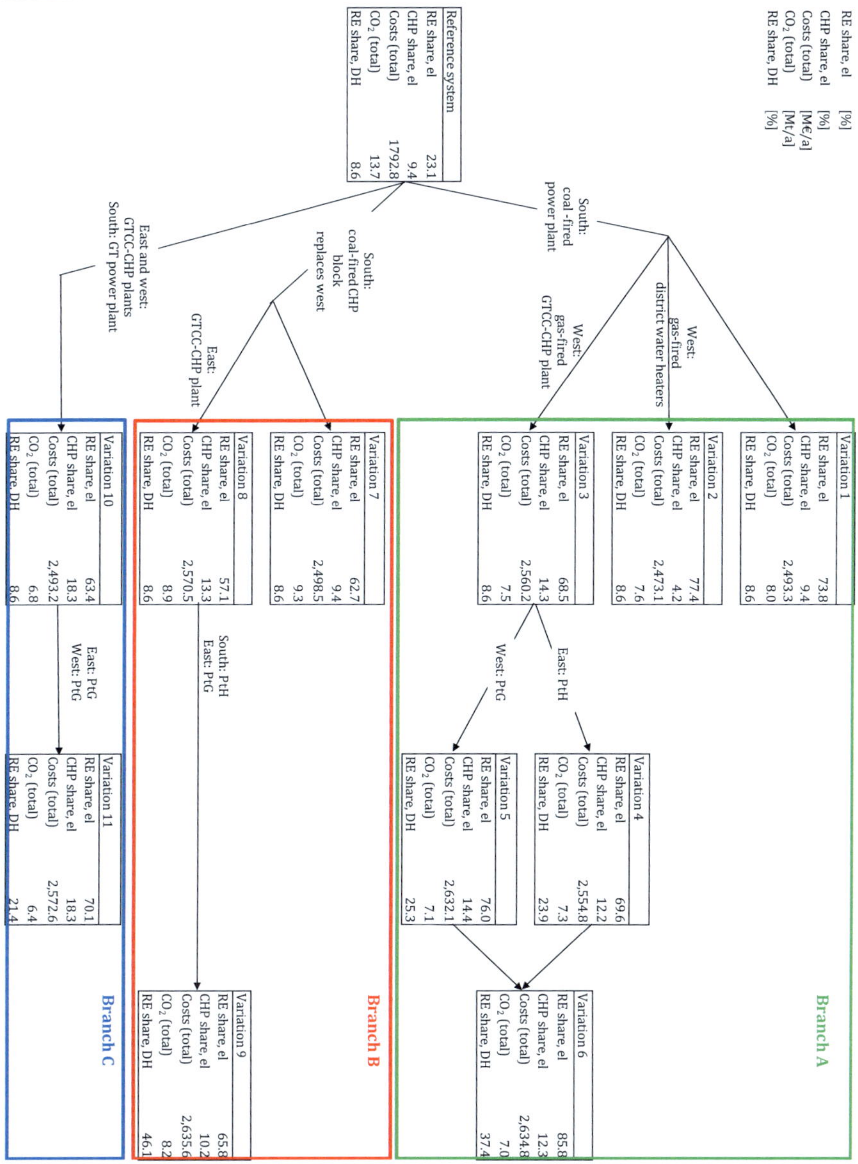

Figure 44 - Summary of result in decision tree format

The definition of the total annual costs (Cost (total)) corresponds to the one presented in equation (24), according to which the costs are calculated individually for every component of the energy park and then cumulated. Fixed costs and variable costs are considered. The costs incurred by the heat producers outside of the large district heating grid are considered as well. The definition of the total annual CO_2 emissions (CO_2 (total)) also corresponds to the one presented in equation (23), according to which the CO_2 emissions are calculated by multiplying the fuel-specific CO_2 emission factors by the fuel consumption of each production unit fired with fossil fuels. In variations where hydrogen is co-fired in the CCGT-CHP plant, the heating input coming from the hydrogen is considered to be free of CO_2 emissions.

As defined in equation (27), the share of RE in district heating consumption (RE share, DH) consists of up to three terms depending on the constitution of the variation's energy park: the first term is the heat coming from the waste-fired heating plant, 50 % of which is considered to be of biogenic nature according to [67]; the second term corresponds to the heat produced by the PtH unit, which is operated only with RE surplus; the third term corresponds to the heat produced by hydrogen co-firing in the CCGT-CHP units, which is calculated according to the description of section 4.2.3.

6.4 Comments on each variation

The reference system is the starting point of this analysis. The share of RE in electricity consumption of 23.1 % reflects the value achieved in Germany in the year 2012 well, which according to [67] was 23.7 %. The share of CHP power in electricity consumption equals 9.4 %. The main goal of the German CHP law of 2015 [85] is to increase the power production from CHP plants to 110 TWh_{el} in 2020 and 120 TWh_{el} in 2025. Considering that in 2012 the power production from CHP plants was 106.5 TWh_{el} [86] the aforementioned goals of the CHP law imply an increase of 3.3 % until 2020 and 12.7 % until 2025 with respect to the year 2012. To reflect this goal of increasing the role of CHP plants in the energy mix, a similar increase of 12.7 % in the share of CHP power in electricity consumption is considered to be satisfactory in this work. Therefore, an increase from 9.4 % in 2012 to more than 10.6 % leads to a satisfactory solution regarding the share of CHP power in electricity consumption. The total annual costs of €1,792.8 million are only an indicative value which serves as reference for the other variations. The calculated total annual CO_2 emissions of 13.7 million tons is consistent with the value published by the Hamburg's Ministry of Urban Development and Environment in [87] for the year 2012. The published value equals 14.2 million tons if the CO_2 emissions produced by the transportation sector, which are not considered in this work, are neglected. The share of RE in district heating consumption of the reference system equals 8.6 %, which accounts for 50 % of the heat produced by the waste incineration plant.

Branch A

In variation 1, which is the starting point of branch A, the share of RE in electricity consumption increases 50.7 %-points with respect to the reference system, to a value of 73.8 %, due to the assumed increase in the RE installed capacity. However, this leads to a considerable amount of RE being curtailed in periods of negative residual load, i.e. periods of local RE surplus. These periods are shown in Figure 16 and Figure 17 and result from an increase in the total RE installed capacity from 1,758 MW_{el} in the reference scenario (2012) to 5,912 MW_{el} in the rest of the variations (2050). The share of CHP power in electricity consumption remains constant at 9.4 % because the CHP plants in this variation remain unchanged with respect to the reference system. Further evident consequences of an increased RE capacity are the higher total annual costs of €2,493.3 million (39.1 % increase with respect to the reference system) and the lower annual CO_2 emissions of 8.0 million tons (decrease of 41.7 % with respect to the reference system). The increase in total annual costs is related to the higher investment costs of RE. The decrease in the annual CO_2 emissions is related to the fact that the positive residual load, which is covered by the coal-fired power plant, is reduced due to the higher RE capacity (compare Figure 14 and Figure 15 with Figure 16 and Figure 17). If the curtailed RE of variation 1 would be utilized in some other form, then the CO_2 emissions could be further reduced. The share of RE in district heating consumption of this variation remains unchanged with respect to the reference system and equals 8.6 %.

In variation 2 an interesting phenomenon can be observed: by replacing the CHP plant at the west site with a gas-fired heating plant, the must-run electricity production of the replaced CHP plant is eliminated. This enables an increase of 3.6 %-points in the share of RE in electricity consumption with respect to variation 1, to a value of 77.4 %. However, this also leads to a reduction in the share of CHP power in electricity consumption of 5.2 %-points with respect to variation 1, to a value of only 4.2 %. In fact this is the variation with the lowest share of CHP in electricity consumption of all. The total annual costs of this variation are 0.8 % lower than the total annual costs of variation 1. The first explanation for this is the fact that a unit with low fuel costs but relatively high investment costs (coal-fired CHP plant) is replaced by a unit with higher fuel costs but low investment costs (gas-fired heating plant) and both effects compensate each other. The CO_2 emissions of this variation remain almost unchanged with respect to variation 1; they just decrease 4.8 %, to 7.6 million tons. The share of RE in district heating consumption of this variation remains unchanged with respect to the reference system and equals 8.6 %.

In variation 3, the share of CHP power in electricity consumption increases 4.9 %-points with respect to variation 1, to a value of 14.3 %. This is due to the presence of the CCGT-CHP plant at the west site, since the CHP coefficient of this plant is higher than that of the coal-fired CHP plant being substituted. The increase in must-run production of the CCGT-CHP plant due to its higher CHP coefficient leads to a reduction of 5.3 %-points in the share of RE in electricity consumption with respect to variation 1, to a value of 68.5 %. The total annual costs increase 2.7 % with respect to variation 1, to €2,560.2 million, due to the usage of natural gas in the CCGT-CHP plant. This increase occurs although the investment costs for the CHP plant at the west site are reduced with respect to variation 1, because the specific investment costs of the CCGT-CHP plant are lower than the specific investment costs of the coal-fired CHP plant (see Table 6 for specific investment costs and Figure 5 for installed capacities). On the contrary, the total annual CO_2 emissions in this variation are reduced by 5.5 % with respect to variation 1 to 7.5 million tons. The share of RE in district heating consumption of variation 3 remains unchanged with respect to the reference system and equals 8.6 %. The following variations of branch A (variations 4, 5 and 6) are based on the system configuration of variation 3.

In variation 4, additionally to the energy park of variation 3, a PtH unit is installed at the east site. As a consequence, the share of RE in electricity consumption reaches a value of 69.6 %, which represents an increase of 1.1 %-points with respect to variation 3. Intuitively, the PtH unit should not affect the share of RE in electricity consumption, but only the share of RE in the district heating consumption. The CHP curtailment effect, which was first presented in section 5.5 is responsible for the aforementioned increase: while operating in backpressure mode, a reduction in the set value of the CHP plant's heat production caused by the usage of the PtH unit leads automatically to a reduction in its must-run electricity production. This leads to a reduction in the share of CHP power in electricity consumption. On the other hand, this also means that more RE can be directly used to cover the system's electricity demand, and the share of RE in electricity consumption increases. Despite the additional investment costs for the PtH unit, the total annual costs decrease 0.2 % with respect to variation 3 to a value of €2,554.8 million because the PtH unit enables savings on the coal consumption of the CHP plant. The CO_2 emissions decrease to a value of 7.3 million tons. This represents a reduction of 3 % with respect to variation 3 and is caused by an increase in the share of RE in electricity and district heating consumption. The share of RE in district heating consumption of this variation reaches a value of 23.9 %, which represents an increase of 15.3 %-points with respect to variation 3. This is due to the presence of the PtH unit, which converts otherwise curtailed RE into heat.

In variation 5, additionally to the energy park of variation 3, an electrolyzer park is installed together with an underground hydrogen storage close to the west site. The hydrogen is produced with otherwise curtailed RE and it is co-fired at the CCGT-CHP plant at the west site. The share of RE in electricity increases in this variation to

76.0 %, a gain of 7.5 %-points with respect to variation 3. The share of RE in district heating consumption increases 16.7 %-points with respect to variation 3, to a value of 25.3 %. These shares are higher than those of variation 3 due to the allocation of the electricity and heat production with hydrogen to the RE share terms. As described in sections 4.2.3 and 5.6, this phenomenon is referred to as the RH heat and power contribution in this work. Furthermore, the hydrogen co-firing in the CCGT-CHP plant leads to lower natural gas consumption and, consequently, to lower annual CO_2 emissions, which decrease 6.1 % with respect to variation 3, to a value of 7.1 million tons. High investment costs for the electrolyzer park and the underground hydrogen storage lead to higher total annual costs, which increase 2.8 % with respect to variation 3, to a value of €2,632.1 million. Under the current assumptions, the natural gas savings do not compensate for the investment costs of the PtG unit.

Variation 6 is a combination of variations 4 and 5. A PtH unit is installed at the east site, which produces heat with otherwise curtailed RE, displacing the heat production of the coal-fired CHP plant. At the west site, a PtGtCHP plant is installed. In this variation, the share of RE in electricity consumption increases 17.3 %-points with respect to variation 3, to a value of 85.8 %, due to the combination of the CHP curtailment effect and the RH power contribution. The share of RE in district heating consumption increases 28.8 %-points with respect to variation 3, to a value of 37.4 %, due to the presence of the PtH unit and the RH heat contribution. In fact, this is the variation with the highest share of RE in electricity consumption of all, and the highest share of RE in district heating consumption of branch A. The total annual costs increase only 0.1 % with respect to variation 5, to a value of €2,634.8 million, making it the most expensive variation of this branch. At the same time, this is the system configuration with the lowest total annual CO_2 emissions of this branch. The CO_2 emissions decrease 7.6 % with respect to variation 3, to a value of 7.0 million tons.

Branch B

The branch B starts with variation 7, in which the CHP plant at the west site is eliminated and its heat production is provided by a larger coal-fired CHP plant located at the south site. Here, the share of RE in electricity consumption decreases to a value of 62.7 % (a reduction of 11.1 %-points with respect to variation 1) because the CHP plant at the south site has a higher must-run production limit than the replaced CHP plant at the west site, due to its larger nominal capacity (see definition of variation 7 in Figure 5). A larger power and electricity output implies a larger steam generator. However, the minimum load limitations of the steam generators (around 25 % of the maximum load) are largely valid for different steam generator sizes. The total annual CO_2 emissions reach a value of 9.3 million tons, an increase of 16.3 % with respect to variation 1, also because of the higher must-run profile of the CHP plant at the south site. The total annual costs are €2,498.5 million, which represent a slight increment of 0.2 % with respect to variation 1. The share of RE in district heating consumption of this variation remains unchanged with respect to the reference system and equals 8.6 %.

Variation 8 builds upon the assumptions of the previous variation, but additionally, the coal-fired CHP plant at the east site is substituted with a CCGT-CHP plant. The resulting share of RE in electricity consumption decreases 5.6 %-points, with respect to variation 7, to a value of 57.1 %. This is because the CHP coefficient of the CCGT-CHP plant is higher than that of the replaced coal-fired CHP plant at the east site. The share of CHP power in electricity consumption increases 3.9 %-points with respect to variation 7, to a value of 13.3 %. This is also because of the higher CHP coefficient of the CCGT-CHP plant. The total annual costs increase 2.9 % with respect to variation 7 to €2,570.5 million mainly because of the higher fuel consumption and fuel costs of the CCGT-CHP plant. Although the change of fuel leads to a reduction of 3.9 % in the total annual CO_2 emissions with respect to variation 7, the 8.9 million tons emitted in this variation are still above the 8.0 million tons of variation 1. The share of RE in district heating consumption of this variation remains unchanged with respect to the reference system and equals 8.6 %.

Variation 9 is an extension of variation 8 with additional PtH and PtG units at the south and east sites. Due to the CHP curtailment effect, the share of RE in electricity consumption increases to a value of 65.8 %, and the share of CHP power in electricity consumption decreases to a value of 10.2 % in this variation. Compared to variation 8, these changes represent an increase of 8.7 %-points and a decrease of 3.1 %-points respectively. The RH power contribution leads also to this increase in the share of RE in electricity consumption. The share of RE in district heating consumption increases 37.5 %-points with respect to variation 8, to a value of 46.1 %, due to the presence of the PtH unit and the RH heat contribution. This is in fact the variation with the highest share of RE in district heating consumption of all. The total CO_2 emissions decrease to a value of 8.2 million tons, which represents a reduction of 7.8 % with respect to variation 8. However, the achieved CO_2 level is still at the level of the variation 1. From a CO_2 emissions perspective, the branch B does not bring any advantage with respect to variation 1. That being said, the total annual costs of €2,635.6 million are 2.5 % higher than the total annual cost obtained in variation 8, making variation 9 the one with the highest costs of this branch. These total annual costs are 5.7 % higher than the total annual cost obtained with variation 1. Therefore, from a CO_2 emissions and costs perspective, variation 9 is not a good alternative to variation 1.

Branch C

Finally, the results from branch C are addressed. In variation 10 the coal-fired power plant at the south site is replaced by a GT power plant and the coal-fired CHP plants are replaced by gas-fired CCGT-CHP plants at the east and west sites. The total annual costs remain basically unchanged with respect to variation 1. This suggests that the increase in costs due to the increased use of natural gas instead of coal compensates the decrease of costs due to the lower specific investment costs of GT and CCGT-CHP plants with respect to coal-fired power and CHP-plants (see Table 6). The CO_2 emissions decrease to a value of 6.8 million tons. This represents a reduction of 15.2 % with respect to variation 1, due to the substitution of coal with natural gas, which has considerably less specific CO_2 emissions than coal. The share of CHP power in electricity consumption increases 8.9 %-points with respect to variation 1, to a value of 18.3 %. This is because both CCGT-CHP plants at the east and west sites have higher CHP coefficients than their coal-fired counterparts. The share of RE in electricity consumption decreases to a value of 63.4 %, which represents a reduction of 10.4 %-points with respect to variation 1 due to the higher CHP coefficient of the CCGT-CHP plants. The share of RE in district heating consumption of this variation remains unchanged with respect to variation 1 and equals 8.6 %. Variation 10 presents a good alternative to variation 1 because its total annual costs remain almost unchanged, while its total CO_2 emissions are reduced.

In variation 11, additionally to variation 10, PtGtCHP plants are installed at the east and west sites. The total annual costs are €2,572.6 million, which represents an increase 3.2 % with respect to variation 10 due to the investment costs of the hydrogen production, storage and transportation system. On the other hand, the CO_2 emissions decrease 5.2 % with respect to variation 10, to a value of 6.4 million tons. This is the variation with the lowest total annual CO_2 emissions of all, and is therefore a good alternative to variation 1. This variation also displays an increase with respect to variation 10 of 6.7 %-points (to a value of 70.1 %) in the share of RE in electricity consumption and an increase of 12.8 %-points (to a value of 21.4 %) in the share of RE in district heating consumption. This is due to the RH power and heat contribution, making it the variation with the highest share of RE in district heating consumption of branch C. Finally, this variation presents a share of CHP power in electricity consumption of 18.3 %, which represents an increase of 8.9 %-points with respect to variation 1, due to the higher CHP coefficients of the CCGT-CHP plants with respect to their coal-fired counterparts.

6.5 Identification of the most promising variations

The simulation results of all system variations can be visualized and analyzed with the help of the diagrams shown in Figure 45, Figure 46 and Figure 47. These illustrate the relations between the total annual costs and CO_2 emissions, between the share of RE in electricity consumption and the share of CHP power in electricity consumption, and between the share of RE in electricity consumption and the share of RE in district heating consumption, respectively.

The process of selecting the best system variation can be regarded as a multi-objective optimization process in which the total annual costs and CO_2 emissions are objectives which should be minimized, and the share of RE and CHP power in electricity consumption, as well as the share of RE in district heating consumption, are objectives which should be maximized. Although no optimization algorithm is used in this work for the simulations, the variations which better satisfy the above mentioned objectives can be identified with the help of these diagrams.

According to Figure 45, increasing the RE production capacity leads to a significant decrease in the total annual CO_2 emissions, but it also leads to an increase in the total annual costs (compare for instance the reference system with variation 1). The increase in RE production capacities which currently takes place worldwide makes evident that society is ready to accept higher costs in order to reduce CO_2 emissions. For this reason, although variations with fewer costs and less CO_2 emissions present an attractive alternative to the current situation, it does not mean that variations which lead to less CO_2 emissions but higher costs should not be considered viable. The results show that all variations within each branch lead to a reduction in the total annual CO_2 emissions, usually coupled with an increase in the total annual costs (compare variations 1 until 6, variation 7 until 9, and variation 10 with 11).

It is also evident that substituting coal-fired units with gas-fired units leads to a significant reduction in the CO_2 emissions. In some cases, the fuel substitution also leads to an increase in total annual costs (compare variation 1 with 3, and variation 7 with 8). In variation 2, the total annual costs decrease slightly with respect to variation 1 (0.8 % reduction) due to the lower investment costs of the gas-fired heater compared to the coal-fired CHP plant at the west site. In contrast, variation 10 leads to a considerable reduction in the total annual CO_2 emissions with respect to variation 1 (15.2 % reduction), without a significant impact on the total annual costs. This can be explained by the lower specific investment costs of GT power plants compared to coal-fired power plants and the fact that, in the 2050 case, the residual load covered by these plants over the year is quite small, which means that the fuel consumption, and therefore the annual fuel costs of the GT power plants, are not that high although the specific fuel costs of natural gas are higher than the specific fuel costs of coal.

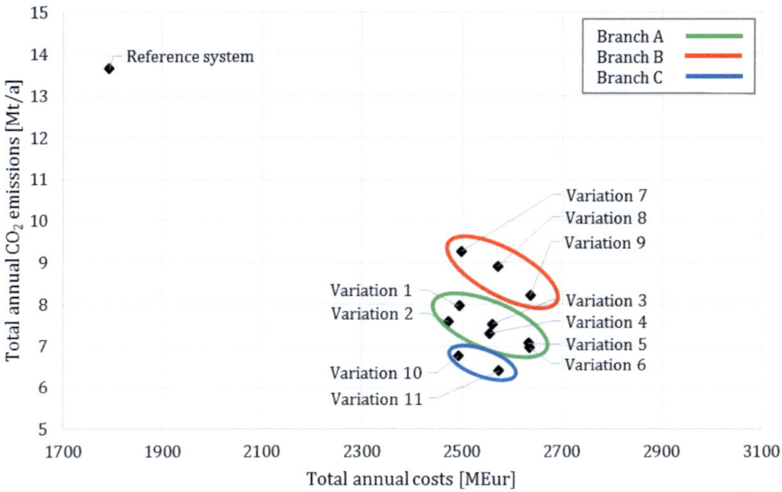

Figure 45 – Total annual costs and CO_2 emissions

By analyzing branch B, it appears that, under the stringent assumption of this work regarding the must-run nature of CHP plants, using heat from the coal-fired plant at the south site leads to an increase in CO_2 emissions without an impact on the total annual costs (compare variation 1 with variation 7). The reason for this increase in the total annual CO_2 emissions is the fact that the CHP plant at the south site has a higher nominal capacity (and therefore a higher must-run production limit) than the replaced CHP plant at the west site, as explained in section 6.4.

The use of PtH units to integrate otherwise curtailed RE in the district heating grid leads to a modest reduction in the total annual CO_2 emissions due to the CHP curtailment effect without a significant effect on the total annual costs (compare variation 4 with 3, and variation 6 with 5). The use of PtGtCHP plants leads to a decrease in the total annual CO_2 emissions and to an increase in the total annual costs (compare variation 5 with 3, and variation 11 with 10).

As a conclusion, variation 10 offers a good relation between total annual costs and CO_2 emissions. However, since the total annual costs and CO_2 emissions are not the only optimization objectives, all variations are now be compared with respect to the share of RE and CHP power in electricity consumption.

Figure 46 shows that an increase in the RE production capacity without altering the must-run production of CHP plants leads to an increase in the share of RE in electricity consumption, while the share of CHP power in electricity consumption remains unchanged (compare variation 1 with the reference system).

The substitution of a coal-fired CHP plant with a gas-fired heating plant allows an increase in the share of RE in electricity consumption, but leads to a decrease in the share of CHP power in electricity consumption (compare variation 2 with 1). Variation 2 displays the lowest share of CHP power in electricity consumption of all variations. The substitution of coal-fired CHP plants with CCGT-CHP plants leads to an increase in the share of CHP power, but to a decrease in the share of RE in electricity consumption due to the higher CHP coefficient of CCGT-CHP plants (compare variation 3 with 1).

PtH units lead to a slight increase in the share of RE, but to a significant decrease in the share of CHP power in electricity consumption due to the CHP curtailment effect (compare variation 4 with 3). Due to the RH power contribution, PtGtCHP plants lead to a significant increase in the share of RE, without negatively affecting the share of CHP power in electricity consumption (compare variation 5 with 3, and variation 10 with 11). Variations 10 and 11 present the highest values regarding the share of CHP in electricity consumption. The combination of the CHP curtailment effect of PtH units and the RH power contribution leads to high shares of RE in electricity consumption (see variation 9 and variation 6).

Variation 6 presents the best performance regarding the share of RE in electricity consumption and satisfies the requirements regarding the share of CHP, defined in section 6.4, by reaching a value above 10.6 %. Variation 10, which was the optimal solution regarding total annual costs and CO_2 emissions, achieves the best values regarding the share of CHP, but presents lower values than variation 1 regarding the share of RE in electricity consumption. In contrast, variation 11, which can be seen as the second best solution regarding costs and CO_2 emissions, maintains a performance similar to variation 1 regarding the share of RE, while keeping the share of CHP in electricity consumption in the same level as variation 10.

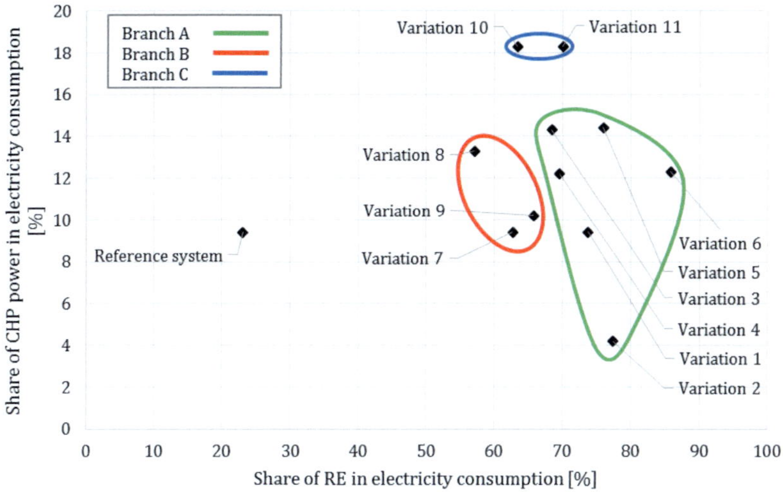

Figure 46 - Share of RE and CHP power in electricity consumption

There seems to be no European or national goal currently regarding the share of RE in district heating consumption. Therefore, this quantity is considered in this work as an additional selection criterion with less weight than the shares of RE and CHP in electricity consumption. As shown in Figure 47, all variations in which no RE integration technologies are implemented maintain the same value regarding the share of RE in district heating consumption, which corresponds to the fraction of the heat output of the waste-fired heating plant that can be considered renewable. PtH units lead to a significant increase in the share of RE in district heating consumption (compare variation 4 with 3). PtGtCHP plants lead to a significant increase in the share of RE in electricity and district heating consumption (compare variation 5 with 3). The combined effects of PtH and PtGtCHP plants lead to the best results regarding the share of RE in electricity and district heating consumptions (see variations 6 and 9). As previously stated, the share of RE in district heating consumption has less weight than the other selection criteria. Although variation 9 presents the highest values regarding this quantity, it fails with respect to its total annual costs and CO_2 emission values. Variation 6 is, as before, the best regarding the share of RE in electricity but it is also very good regarding the share of RE in district heating consumption. However, its total annual costs and CO_2 emission values are higher than the variations in branch C. Considering these facts, variation 11 presents a good compromise between share of RE in district heating consumption, total annual costs and total annual CO_2 emissions.

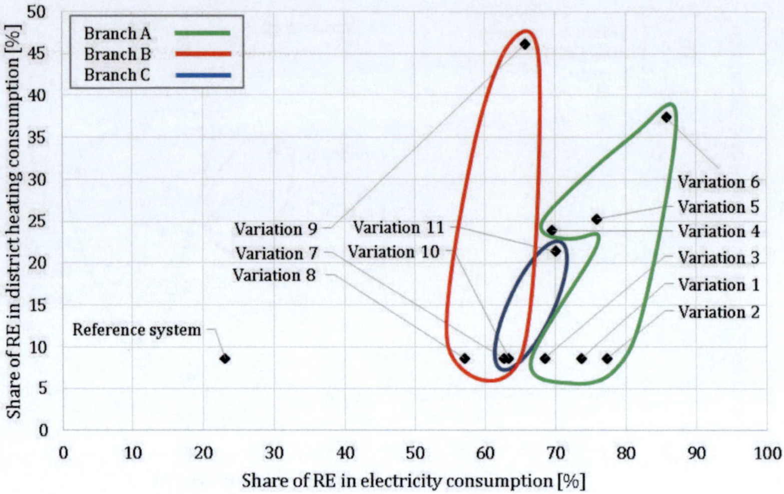

Figure 47 – Share of RE in electricity and district heating consumption

In conclusion, two variations can be selected, depending on which performance indicators are given more weight. If more weight is given to the total CO_2 emissions than to the share of RE in electricity consumption, then variation 11 is suggested as the best choice, because it presents the lowest total annual CO_2 emissions, together with a reasonable increase in total annual costs, as well as satisfactory shares of RE and CHP power in the electricity consumption. If instead, more weight is given to the share of RE in electricity than to the total annual costs and CO_2 emissions, then variation 6 is suggested as the best choice, because it presents the highest share of RE in electricity consumption and reasonable reductions in the total annual CO_2 emissions. Although this variation has one of the highest values regarding the total annual costs, the increase of these costs with respect to variation 1 is moderate. For these reasons, variations 11 and 6 are identified as the most promising system alternatives considering the assumptions taken in this work.

6.6 Comments on security of supply

As described in section 3.1, the energy system in this work is considered as an energy island able to satisfy its own energy demands at all times. To achieve this, the installed power production capacity should be at least as large as the system's peak electricity demand. A security of supply factor is defined, as indicator of the system's ability to cover its own power demand

$$Security\ of\ supply\ factor\ (dispatchable) = \frac{\sum_{i=1}^{n} P_{nom,dispatchable,i}}{P_{demand,max}} \quad (34)$$

where n is the number of dispatchable power producers. Conventional power plants, CHP plants, and biomass power plants are considered in this term. Non-dispatchable power plants such as PV, onshore wind, offshore wind, run-off hydro, and pumped hydro storage are not considered. $P_{nom,dispatchable,i}$ is the nominal production capacity of the ith dispatchable power producer. $P_{demand,max}$ is the system's peak electricity demand, which in this case equals 2,011 MW_{el}.

The security of supply factors of all system variations shown in Table 8 are calculated using these definitions and the nominal production capacity values, presented in Figure 5. By analyzing these results, it becomes evident that the system's dispatchable components are able to completely cover the system's peak power demand, even if the non-dispatchable RE plants were not able to produce any power. This fact is in line with the energy island approach followed in this work.

Table 8 – Security of supply factors of the system variations

	Security of supply factor (dispatchable)
Reference system	1.1
Variation 1	1.3
Variation 2	1.1
Variation 3	1.4
Variation 4	1.4
Variation 5	1.4
Variation 6	1.4
Variation 7	1.1
Variation 8	1.3
Variation 9	1.3
Variation 10	1.5
Variation 11	1.5

Although not analyzed in this work, removing the restriction of a dispatchable security of supply factor above 1 would enable a reduction in the system's total annual costs due to a reduction in the total investment costs. Technically viable solutions with lower security of supply factors could be achieved by carefully dimensioning the system's power production capacity while considering cases with low power output from non-dispatchable RE sources. However, the costs reduction achieved in this way would come together with higher risks regarding the system's security of supply (see example of the "dark doldrums" in section 2.1). The relation between the system's total annual costs, total annual CO_2 emissions, and security of supply is hereby confirmed.

7 FINAL REMARKS

High RE power production capacities together with insufficient power transmission capacities can lead to regional balancing issues between power production and demand. In periods of time in which the RE production exceeds the regional demand, congestion management measures such as re-dispatch and countertrading are needed. Nowadays, the curtailment of RE is one of the main instruments to restore the regional balance. At the same time, for periods of time in which the RE production lays below the demand, the residual load must be covered with other energy sources in an economically and environmentally favorable way.

In this work, several variations of an urban energy system were defined, modeled, simulated and analyzed in an attempt to increase the share of RE in electricity and district heating consumption, as well as the share of CHP power in electricity consumption. At the same time, a decrease in the total annual CO_2 emissions of the system was pursued and the total annual costs of each of the system variations were calculated.

In a first step, the system boundaries, the input data and the system variations were defined. Subsequently, the modeled system and its components were described. A description of the energy producers, the RE integration technologies and the models in charge of the superordinate unit dispatch was included. The simulation results of each system variation were first presented in form of weekly and yearly production profiles for power and heat. Sankey diagrams were then used to illustrate the annual energy and CO_2 emission flows, which resulted from the simulations. Finally, annual performance indicators were defined and used to compare the annual results and identify promising system variations.

The results show that fuel substitution, i.e. replacing a coal-fired CHP plant with a gas-fired CCGT-CHP plant, leads to significant CO_2 reductions. But, in the European fuel price setting assumed in this work, this substitution leads also to higher annual costs because of the higher price of natural gas with respect to coal.

The results also suggest that adding a PtH unit can contribute to a slight reduction in the total CO_2 emissions without a significant effect on the total annual costs. Substituting heat from fossil fuels with heat from RE surplus leads to this reduction.

PtGtCHP plants have a considerable impact on reducing the total CO_2 emissions, but have a significant negative impact in the total annual costs. Despite the fact that the natural gas consumption of these plants is reduced because of the RH co-firing, the high investment costs of the electrolyzers and the underground hydrogen storage overweight the fuel costs savings. In the future, a change in the natural gas prices or in the investment costs of electrolyzers could lead to other conclusions regarding the economics of PtGtCHP plants.

Regarding the sector coupling, the simulations suggest the existence of two interesting phenomena. In this work, these are referred to as the CHP curtailment effect and the RH power and heat contribution. The CHP curtailment effect appears when a PtH unit takes over the heat production in periods of high RE production, displacing the heat production from a CHP unit. For CHP units operating in backpressure mode, a reduction in the heat production leads to a reduction in the power production, which indirectly allows the direct utilization of more RE. The RH power and heat contribution occurs in system variations with PtGtCHP plants. This effect consists of an increase in the share of RE in electricity and district heating consumption because the power and heat produced with RH can be considered renewable.

The simulations show the conflict between the goals of increasing the share of RE and CHP power in electricity consumption: RE curtailment increases if the installed CHP capacity increases due to the must-run characteristics of CHP plants assumed in this work. PtH units lead to a light increase in the share of RE in electricity consumption due to the CHP curtailment effect, but also lead to a reduction in the share of CHP power in electricity consumption. If the must-run characteristic of the CHP plants would be removed, these could be turned off when the RE surplus is enough to cover the plants' power and heat production. However, in this case the CHP curtailment effect would be higher and the share of CHP power in electricity consumption would be lower. Only the PtGtCHP plants allow a simultaneous increase in the share of RE and in the share of CHP power in electricity consumption, as well as an increase in the share of RE in district heating consumption.

Two variations were identified as interesting alternatives in a future urban energy system with high shares of RE and a centralized energy park. The results showed that, in variation 11, the total annual CO_2 emissions can be reduced by 53.1 % with respect to the reference system. However, this would lead to an increase of 43.5 % in the total annual costs. If more weight is given to increasing the share of RE in electricity consumption, then variation 6 is the best choice, because it achieves a share of RE in electricity consumption of 85.8 %. This comes, however, with an increase of 47.0 % in the total annual costs with respect to the reference system. A reduction in the system's power production capacity could offer a way to diminish the total annual costs. But this reduction would imply higher risks regarding the system's security of supply.

From a technological perspective, achieving a transformation in future urban energy systems towards lower CO_2 emissions and higher shares of RE in electricity and heat consumption seems possible. However, society must decide if it is willing to pay the higher costs associated with this quest. Costs reductions of technologically viable system configurations could be possible either by future reductions in the system's specific investment costs and specific fuel costs, or by reductions in the system's security of supply.

References

[1] GERMAN BUNDESTAG. *Erneuerbare-Energien-Gesetz vom 21. Juli 2014 (BGBl. I S. 1066), das durch Artikel 2 des Gesetzes vom 22. Dezember 2016 (BGBl. I S. 3106) geändert worden ist* [online]. Bundesministerium der Justiz und für Verbraucherschutz, 2017 [viewed 20 March 2017]. Available from: https://www.gesetze-im-internet.de/bundesrecht/eeg_2014/gesamt.pdf

[2] FRAUNHOFER ISE. *Installierte Netto-Leistung zur Stromerzeugung in Deutschland in 2016* [online]. Energy-Charts.de, 2017 [viewed 17 February 2017]. Available from: https://www.energy-charts.de/power_inst_de.htm

[3] ANDRESEN, L., DUBUCQ, P., PENICHE GARCIA, R., ACKERMANN, G., KATHER, A. and SCHMITZ, G. Status of the TransiEnt Library : Transient simulation of coupled energy networks with high share of renewable energy. In: FRITZSON, Peter and ELMQVIST, Hilding (eds.), *Proceedings of the 11th International Modelica Conference, Versailles, France, September 21-23, 2015* [online]. Linköping University Electronic Press, 2015. p. 695–705 [viewed 20 March 2017]. Available from: http://www.ep.liu.se/ecp/118/ecp15118.pdf

[4] SYED, A., AHMAD, S., BIALOWAS, A., RICHARDSON, E., NOWAKOWSKI, D. and NICHOLSON, P. *Asia-Pacific Renewable Energy Assessment* [online]. Bureau of Resources and Energy Economics (BREE), 2014 [viewed 20 March 2017]. Available from: https://industry.gov.au/Office-of-the-Chief-Economist/Publications/Documents/aprea/20140703.pdf

[5] SCHUMAN, S. and LIN, A. China's Renewable Energy Law and its impact on renewable power in China: Progress, challenges and recommendations for improving implementation. *Energy Policy* [online]. December 2012, Vol. 51, p. 89–109 [viewed 1 December 2014]. DOI10.1016/j.enpol.2012.06.066. Available from: http://linkinghub.elsevier.com/retrieve/pii/S0301421512006660

[6] BLARKE, M.B. and JENKINS, B.M. SuperGrid or SmartGrid: Competing strategies for large-scale integration of intermittent renewables? *Energy Policy* [online]. April 2013, Vol.58, p.381–390 [viewed 20 March 2017]. DOI10.1016/j.enpol.2013.03.039. Available from: http://dx.doi.org/10.1016/j.enpol.2013.03.039

[7] MUNKSGAARD, J. and MORTHORST, P.E. Wind power in the Danish liberalised power market-Policy measures, price impact and investor incentives. *Energy Policy* [online]. August 2008, Vol. 36, p. 3940–3947 [viewed 20 March 2017]. DOI10.1016/j.enpol.2008.07.024. Available from: http://www.sciencedirect.com/science/article/pii/S0301421508003649

[8] CWE MC PROJECT. *A report for the regulators of the Central West European (CWE) region on the final design of the market coupling solution in the region* [online]. epexspot.com, 2010 [viewed 20 March 2017]. Available from: http://static.epexspot.com/document/7616/01_CWE_ATC_MC_project_documentation.pdf

[9] GERMAN BUNDESTAG. *Energiewirtschaftsgesetz vom 7. Juli 2005 (BGBl. I S. 1970, 3621), das durch Artikel 1 des Gesetzes vom 27. Januar 2017 (BGBl. I S. 130) geändert worden ist* [online]. Bundesministerium der Justiz und für Verbraucherschutz, 2017 [viewed 20 March 2017]. Available from: https://www.gesetze-im-internet.de/bundesrecht/enwg_2005/gesamt.pdf

[10] WETZEL, D. *"Elon" und "Felix" kosten Verbraucher Millionen* [online]. Welt.de, 2015 [viewed 20 March 2017]. Available from: https://www.welt.de/wirtschaft/energie/article136300154/Elon-und-Felix-kosten-Verbraucher-Millionen.html

[11] HAESELER, S. *Stürme ELON und FELIX in Deutschland vom 9. bis 11. Januar 2015* [online]. Deutscher Wetterdienst (DWD), 2015 [viewed 20 March 2017]. Available from: https://www.dwd.de/DE/leistungen/besondereereignisse/stuerme/20150113_elon_felix_deutschland.pdf

[12] TRANSNETBW GMBH. *Übertragungsnetzbetreiber ziehen Fazit zum Winter 2011 / 2012 : Versorgungssicherheit durch große Anstrengungen gewährleistet* [online]. TransnetBW GmbH, 2012 [viewed 20 March 2017]. Available from: https://www.transnetbw.de/uploads/2013-04-26-10-40-42.pdf

[13] WALKER, T. *What happens with German renewables in the dead of winter?* [online]. DW.com, 2017 [viewed 17 February 2017]. Available from: http://www.dw.com/en/what-happens-with-german-renewables-in-the-dead-of-winter/a-37462540

[14] WETZEL, D. *Die „Dunkelflaute" bringt Deutschlands Stromversorgung ans Limit* [online]. Welt.de, 2017 [viewed 17 February 2017]. Available from: https://www.welt.de/wirtschaft/article161831272/Die-Dunkelflaute-bringt-Deutschlands-Stromversorgung-ans-Limit.html

[15] WINTER, W. *European Wind Integration Study (EWIS) - Final report* [online]. ENTSO-E, 2010 [viewed 20 March 2017]. Available from: http://www.wind-integration.eu/downloads/library/EWIS_Final_Report.pdf

[16] STEINKE, F., WOLFRUM, P. and HOFFMANN, C. Grid vs. storage in a 100% renewable Europe. *Renewable Energy* [online]. September 2013, Vol. 50, p. 826–832 [viewed 20 March 2017]. DOI 10.1016/j.renene.2012.07.044. Available from: http://dx.doi.org/10.1016/j.renene.2012.07.044

[17] KOHLER, S., AGRICOLA, A.-C. and SEIDL, H. *DENA-Netzstudie II - Integration erneuerbarer Energien in die deutsche Stromversorgung im Zeitraum 2015-2020 mit Ausblick 2025* [online]. Deutschen Energie-Agentur GmbH, 2010 [viewed 20 March 2017]. Available from: https://www.dena.de/fileadmin/user_upload/Download/Dokumente/Studien__Umfragen/Endbericht_dena-Netzstudie_II.PDF

[18] GE ENERGY. *Western Wind and Solar Integration Study* [online]. National Renewable Energy Laboratory, 2010 [viewed 20 March 2017]. Available from: http://www.nrel.gov/docs/fy10osti/47434.pdf

[19] LEW, D., BRINKMAN, G., IBANEZ, E., FLORITA, A., HEANEY, M., HODGE, B., HUMMON, M., STARK, G., KING, J., LEFTON, S.A., KUMAR, N., AGAN, D., JORDAN, G. and VENKATARAMAN, S. *The Western Wind and Solar Integration Study Phase*

2 [online]. National Renewable Energy Laboratory, 2013 Available from: http://www.nrel.gov/docs/fy13osti/55588.pdf

[20] ENERNEX CORPORATION. *Eastern Wind Integration and Transmission Study* [online]. National Renewable Energy Laboratory, 2011 [viewed 20 March 2017]. Available from: http://www.nrel.gov/docs/fy11osti/47078.pdf

[21] MA, T., OSTERGAARD, P.A., LUND, H., YANG, H. and LU, L. An energy system model for Hong Kong in 2020. *Energy* [online]. March 2014, Vol. 68, p. 301–310 [viewed 20 March 2017]. DOI 10.1016/j.energy.2014.02.096. Available from: http://dx.doi.org/10.1016/j.energy.2014.02.096

[22] RICHTER, S. *Entwicklung einer Methode zur integralen Beschreibung und Optimierung urbaner Energiesysteme. Erste Anwendung am Beispiel Augsburg* [online]. Universität Augsburg, 2004. [viewed 20 March 2017]. Available from: https://opus.bibliothek.uni-augsburg.de/opus4/files/25/dissertation_s_richter_screen.pdf

[23] GRÖGER, M., BERKING, D., LUDWIG, A. and BRUCKNER, T. Kraft-Wärme-Kopplung und Solarthermie : Konkurrenzeffekte beim integrierten Einsatz. In: *Proceedings der 8. Internationale Energiewirtschaftstagung an der TU Wien* [online]. TU Wien, 2013. [viewed 10 October 2013]. Available from: http://eeg.tuwien.ac.at/eeg.tuwien.ac.at_pages/events/iewt/iewt2013/uploads/

[24] BEGLUK, S., BOXLEITNER, M., SCHLAGER, R., HEIMBERGER, M., MAIER, C. and GAWLIK, W. SYMBIOSE und Speicherfähigkeit von dezentralen Hybridsystemen. In: *Proceedings der 8. Internationale Energiewirtschaftstagung an der TU Wien* [online]. TU Wien, 2013. [viewed 20 March 2017]. ISBN 4315880137019. Available from: http:// www.ea.tuwien.ac.at/fileadmin/t/ea/projekte/Symbiose/P_213_Begluk_Sabina_12-Feb-2013_14_40.pdf

[25] PETER, S. and LUTZENBERGER, A. *Insel Stromstudie Hamburg-Wilhelmsburg* [online]. Kasseburg: IBA Hamburg GmbH, 2013 [viewed 20 March 2017]. Available from: http://www.iba-hamburg.de/fileadmin/Mediathek/ Whitepaper/Stromstudie_IBA_Hamburg_klein.pdf

[26] PEHNT, M., FRANKE, B., HERTLE, H., KAUERTZ, B., OTTER, P., GROSCURTH, H.-M., BOSSMAN, T. and KASTEN, P. *Das Steinkohle-Kraftwerk Hamburg Moorburg und seine Alternativen* [online]. Institut für Energie- und Umweltforschung GmbH, 2007 [viewed 20 March 2017]. Available from: http://fb4.ifeu.de/ energie/pdf/IFEU Arrhenius _2007_ - BUND Alternativkraftwerk final V3.pdf

[27] GROSCURTH, H.-M., BODE, S. and KÜHN, I. *Basisgutachten zum Masterplan Klimaschutz für Hamburg* [online]. arrhenius Institut für Energie- und Klimapolitik, 2010 [viewed 20 March 2017]. Available from: http:// www.hamburg.de/contentblob/4312988/d35ac390ff234478e818023286d2a2b4/data/basisgutachten-masterplan-klimaschutz.pdf

[28] ZANDER, W., MICHELS, A., BARTELT, M., HEIMES, K. and DONNER, O. *Erstellung einer Expertise zur Hamburger Fernwärmeversorgung ; Handlungsalternativen*

für das Kohlekraftwerk in Wedel [online]. BET Büro für Energiewirtschaft und technische Planung GmbH, 2015 [viewed 20 March 2017]. Available from: http://www.hamburg.de/content blob/4616726/a45eb664dbe77aecb39356b246498d14/data/endbericht-gutachten-wedel.pdf

[29] 50HERTZ TRANSMISSION GMBH, AMPRION GMBH, TENNET TSO GMBH and TRANSNETBW GMBH. *Netzentwicklungsplan Strom 2014 - Zweiter Entwurf der Übertragungsnetzbetreiber* [online]. 50Hertz Transmission GmbH, Amprion GmbH, TenneT TSO GmbH, TransnetBW GmbH, 2014 [viewed 20 March 2017]. Available from: http://www.netzentwicklungsplan.de/NEP_2014_2_Entwurf_Teil1.pdf

[30] BÜRGERSCHAFT DER FREIEN UND HANSESTADT HAMBURG. *Masterplan Klimaschutz - Zielsetzung, Inhalt und Umsetzung* [online]. Hamburg.de, 2013 [viewed 20 March 2017]. Available from: http://www.hamburg.de/content blob/4050236/data/masterplan-klimaschutz.pdf

[31] DONNER, O. and BRÜHL, S. *Status Quo und Potentialanalyse von Speichertechnologien, Energieinifrastrukturen und Lastmanagement für Unternehmen in der Metropolregion Hamburg* [online]. Hamburg: Erneuerbare Energien Hamburg Clusteragentur GmbH, 2015 [viewed 20 March 2017]. Available from: www.erneuerbare-energien-hamburg.de/en/downloads.html?page_c27=2&file=files/eehh-website/upload/eehh/general/download s/public/Gutachten/Studie Status Quo und Potentialanalyse/20150218-EEHH-Studie_Energiespeicher_2015-Webfassung.pdf

[32] CZIESLA, F., HENCKE, E.-G., KATHER, A., KELLER, D. and RUKES, B. *Statusreport 2013: Fossil befeuerte Großkraftwerke in Deutschland - Stand, Tendenzen, Schlussfolgerungen* [online]. VDI, 2013 [viewed 20 March 2017]. Available from: https://m.vdi.de/uploads/media/3544_BRO_TW_GEU_Statusreport_Fossil_befeuerte_Grosskraftwerke.pdf

[33] OEDING, D. and OSWALD, B.R. *Elektrische Kraftwerke und Netze* [online]. Berlin, Heidelberg: Springer Berlin Heidelberg, 2011. [viewed 14 August 2013]. ISBN9783642192456. Available from: http://www.springerlink.com/index/10.1007/978-3-642-19246-3

[34] E.ON SE. *E.ON continuing to expand CCGT power generation* [online]. Eon.com, 2011 [viewed 5 August 2016]. Available from: http://www.eon.com/en/media/news/press-releases/2011/6/27/e-dot-on-continuing-to-expand-ccgt-power-generation.html

[35] BRAUNER, G., GLAUNSINGER, W., BOFINGER, S., JOHN, M., MAGIN, W., PYC, I., SCHÜLER, S., SCHULZ, S., SCHWING, U., SEYDEL, P., STEINKE, F., FRICK, A., STAMATELOPOULOS, G.-N. and SCHWING, U. *Erneuerbare Energie braucht flexible Kraftwerke - Szenarien bis 2020*. Frankfurt am MaIn: Energietechnische Gesellschaft im VDE, 2012

[36] HAMBURGISCHE ELECTRICITÄTS-WERKE AG. *Heizwerk Hafen - Fernwärme für Hamburg - Energie für eine saubere Zukunft*. Hamburg: Hamburgische Electricitäts-Werke AG, 1996

[37] CERBE, A. *Prozessnahe Einsatzoptimierung mit BoFiT unter Berücksichtigung der Netzrestriktionen* [online]. Hamburg: Hamburgische Electricitäts-Werke AG, 2002 [viewed 9 April 2014]. Available from: http://edok01.tib.uni-hannover.de/edoks/e01fb02/357623436.pdf

[38] STEIN, A. and EHRET, A. KA26 Combined Cycle and CHP Meeting the Requirements of the European Markets. In: *Power-Gen Europe, Cologne, 04. June 2014* [online]. Cologne: ALSTOM Power AG, Rhein Energie AG, 2014. [viewed 20 April 2017]. Available from: http://pennwell.sds06.websds.net/2014/cologne/pge/slideshows/T3S3050-slides.pdf

[39] ERKER, M. *Innovationskraftwerk Wedel* [online]. Vattenfall Wärme Hamburg GmbH, 2013 [viewed 20 March 2017]. Available from: http://www.egeb.de/fileadmin/Dokumente/Foren/130524_Energie_Erker_Innovationskraftwerk_Wedel.pdf

[40] ZIEMS, C., MEINKE, S., NOCKE, J., WEBER, H. and HASSEL, E. *Kraftwerksbetrieb bei Einspeisung von Windparks und Photovoltaikanlagen* [online]. Rostock: Universität Rostock, VGB-PowerTech, 2012 [viewed 20 March 2017]. Available from: https://www.vgb.org/vgbmultimedia/333_Abschlussbericht-p-5968.pdf

[41] BRAUNER, G., GLAUNSINGER, W., BOFINGER, S., JOHN, M., MAGIN, W., PYC, I., SCHÜLER, S., SCHULZ, S., SCHWING, U., SEYDEL, P. and STEINKE, F. *Erneuerbare Energie braucht flexible Kraftwerke - Szenarien bis 2020*. Verband der Elektrotechnik Elektronik Informationstechnik e. V., 2012

[42] ANDERLOHR, T. and GRASSMAN, A. Flexibilisiertung der Betriebsweise von Heizkraftwerke durch Wärmespeicher und Elektrokessel. In: BECKMANN, Michael and HURTADO, Antonio (eds.), *Kraftwerkstechnik 2014 - Strategien, Anlagentechnik und Betrieb*. Freiberg: SAXONIA Standortentwicklungs- und -verwaltungsgesellschaft mbH, 2014. p. 833–845.

[43] GERHARDT, N., RICHTS, C. and HOCHLOFF, P. *Power-to-Heat zur Integration von ansonsten abgeregeltem Strom aus Erneuerbaren Energien abgeregeltem Strom aus Erneuerbaren Energien - Anhang* [online]. BerlIn: Agora Energiewende, 2014 [viewed 20 March 2017]. Available from: https://www.agora-energiewende.de/fileadmin/Projekte/2013/power-to-heat/Agora_PtH_Anhang_WEB.pdf

[44] SMOLINKA, T., GÜNTHER, M. and GARCHE, J. *Stand und Entwicklungspotenzial der Wasserelektrolyse zur Herstellung von Wasserstoff aus regenerativen Energien* [online]. Fraunhofer ISE, FCBAT, 2011 [viewed 20 March 2017]. Available from: http://www.hs-ansbach.de/uploads/tx_nxlinks/NOW-Studie-Wasserelektrolyse-2011.pdf

[45] SIEMENS AG. *SILYZER 200 (PEM electrolysis system) driving convergence between energy and industrial markets* [online]. Siemens AG, 2015 [viewed 14 February 2017]. Available from: https://www.industry.siemens.com/topics/global/en/pem-electrolyzer/silyzer/Documents/silyzer-200-en_v1.3_InternetVersion.pdf

[46] PITSCHAK, B. *Hydrogenics Multi MW PEM Elektrolyzer - a builiding block in the Energiewende* [online]. Hydrogenics GmbH, 2014 [viewed 20 March 2017].

Available from: http://www.japantag-duesseldorf-nrw.de/fileadmin/pdf/wirtschaftstag/Wirtschaftstag_Praesentationen/Pitschak_Hydrogenics_Japantag_2014.pdf

[47] ITM POWER. *100MW electrolyser plant designs to be launched at Hannover* [online]. ITM Power PLC, 2016 [viewed 14 February 2017]. Available from: http://www.itm-power.com/news-item/100mw-electrolyser-plant-designs-to-be-launched-at-hannover

[48] STOLZENBURG, K., HAMELMANN, R., WIETSCHEL, M., GENOESE, F., MICHAELIS, J., LEHMANN, J., MIEGE, A., KRAUSE, S., SPONHOLZ, C., DONADEI, S., CROTOGINO, F., ACHT, A. and HORVATH, P.-L. *Integration von Wind-Wasserstoff-Systemen in das Energiesystem Abschlussbericht* [online]. PLANET Planungsgruppe Energie und Technik GbR, Fachhochschule Lübeck PROJEKT-GMBH, Fraunhofer ISI, Institut für Energie und Umwelt e. V. an der Fachhochschule Stralsund, KBB Underground Technologies GmbH, NOW GmbH, 2014 [viewed 20 March 2017]. Available from: http://www.planet-energie.de/de/media/Abschlussbericht_Integration_von_Wind_Wasserstoff_Systemen_in_das_Energiesystem.pdf

[49] GAHLEITNER, G. Hydrogen from renewable electricity: An international review of power-to-gas pilot plants for stationary applications. *International Journal of Hydrogen Energy* [online]. 2013, Vol. 38, no. 5, p. 2039–2061 [viewed 20 March 2017]. DOI 10.1016/j.ijhydene.2012.12.010. Available from: http://dx.doi.org/10.1016/j.ijhydene.2012.12.010

[50] SCHOOF, R. *Power-to-Gas Anlage Hamburg Reitbrook Flexibilität und Schnittstellen* [online]. Gießen: Uniper Energy Storage GmbH, 2016 [viewed 20 March 2017]. Available from: https://www.energieland.hessen.de/BFEH/giessen/Rene_Schoof_Power-to-Gas-Anlage_Hamburg_Reitbrook.pdf

[51] RIEKE, S. Erste industrielle Power-to-Gas-Anlage mit 6 Megawatt. *gwf-Gas, Erdgas* [online]. 2013, Vol. September, p. 660–664 [viewed 20 March 2017]. Available from: https://www.di-verlag.de/media/content/gwf-GE/gwf_Gas_9_13/gwf-GE_09_2013_fb_Rieke.pdf?xaf26a=7607be2af411325a0ddff83247813f87

[52] FLEIGE, M. *Direkte Methanisierung von CO_2 aus dem Rauchgas konventioneller Kraftwerke*. Springer Spektrum, 2015. ISBN 9783658092245.

[53] MÜLLER-SYRING, G., HENEL, M., KÖPPEL, W., MLAKER, H., STERNER, M. and HÖCHER, T. *Studie Entwicklung von modularen Konzepten zur Erzeugung , Speicherung und Einspeisung von Wasserstoff und Methan ins Erdgasnetz* [online]. DVGW Deutscher Verein des Gas- und Wasserfaches e. V., 2013 [viewed 20 March 2017]. Available from: https://www.dvgw.de/medien/dvgw/leistungen/forschung/berichte/g1_07_10.pdf

[54] EICHLSEDER, H. and KLELL, M. *Wasserstoff in der Fahrzeugtechnik* [online]. Wiesbaden: Vieweg+Teubner Verlag, 2012. [viewed 9 August 2013]. ISBN 978-3-8348-1754-9. Available from: http://www.springerlink.com/index/10.1007/978-3-8348-2196-6

[55] KRÄHLING, K. *Personal communication. 19 March 2017.*

[56] STEWARD, D., SAUR, G., PENEV, M. and RAMSDEN, T. *Lifecycle Cost Analysis of Hydrogen Versus Other Technologies for Electrical Energy Storage* [online]. National Renewable Energy Laboratory, 2009 [viewed 20 March 2017]. Available from: http://www.nrel.gov/docs/fy10osti/46719.pdf

[57] BROWER, M., PETERSEN, E.L., METCALFE, W., CURRAN, H.J., FÜRI, M., BOURQUE, G., ALURI, N. and GÜTHE, F. Ignition Delay Time and Laminar Flame Speed Calculations for Natural Gas/Hydrogen Blends at Elevated Pressures. *Journal of Engineering for Gas Turbines and Power* [online]. 2013, Vol. 135, p. 21504 [viewed 20 March 2017]. DOI 10.1115/1.4007763. Available from: http://dx.doi.org/10.1115/1.4007763

[58] KNAPP, K. Experience with variable natural gas compositions and use of hydrogeneous fuels in a 300 MW gas turbine type GT26. In: *VGB Conference, Gas Turbines and Operation of Gas Turbines 2013*. Friedrichshafen: ALSTOM, 2013.

[59] MARX, P., LIEBAU, M., GASSER-PAGANI, B. and STEVENS, M. Fuel Flexibility Capabilities of Alstom 's GT26 Gas Turbine. In: *Power-Gen Europe, Vienna, 4-6 June 2013* [online]. Vienna: ALSTOM, 2013. [viewed 20 March 2017]. Available from: http://pennwell.sds06.websds.net/2013/vienna/pge/papers/T3S7O3-paper.pdf

[60] WU, J., BROWN, P., DIAKUNCHAK, I., GULATI, A., GENERATION, S.P., LENZE, M. and KOESTLIN, B. Advanced Gas Turbine Combustion System Development for High Hydrogen Fuels. *Proceedings of GT2007, ASME Turbo Expo 2007: Power for Land, Sea and Air May 14-17, 2007, Montreal, Canada* [online]. 2007, Available from: https://www.energy.siemens.com/hq/pool/hq/power-generation/power-plants/integrated-gasification-combinedcycle/Advanced-gas-turbine-combustion.pdf

[61] LAMMEL, O., SCHÜTZ, H., SCHMITZ, G., LÜCKERATH, R., STÖHR, M., NOLL, B., AIGNER, M., HASE, M. and KREBS, W. FLOX® Combustion at High Power Density and High Flame Temperatures. *Journal of Engineering for Gas Turbines and Power* [online]. 2010, Vol. 132, p. 121503 [viewed 20 March 2017]. DOI10.1115/1.4001825. Available from: http://gasturbinespower.asme digitalcollection.asme.org/article.aspx?articleid=1429076

[62] STROMNETZ HAMBURG GMBH. *Jahreshöchstlast der Netzlast und Lastverlauf* [online]. stromnetz-hamburg.de, 2014 [viewed 29 September 2014]. Available from: http://www.stromnetz-hamburg.de/de/jahreshoechstlast-der-netzlast.htm

[63] BEECKEN, J., FEUERRIEGEL, S. and STAPF, K. *Maximierung der Kraftproduktion durch optimale Nutzung des Energiebedarfes im Fernwärmenetz (KWK-Optimierung) - Band 5: Teilbericht Auswertung* [online]. Hamburg: Vattenfall Europe Hamburg AG, 3S Consult GmbH, Franke+Pahl GmbH, 2007 [viewed 2 August 2013]. Available from: https://www.tib.eu/suchen/id/TIBKAT:566424452/

[64] DEUTSCHER WETTERDIENST. *WebWerdis - ausgewählte meteorologische Daten und Produkte online* [online]. DWD.de, 2015 [viewed 20 March 2017]. Available from: http://www.dwd.de/DE/leistungen/webwerdis/webwerdis.html

[65] BRUNNENGRÄBER, B. and LOGA, T. *Jahresdauerlinien für Niedrigenergiesiedlungen - Gemessene Tagesganglinien als Grundlage für die Auslegung von Blockheizkraftwerken* [online]. Darmstadt: Institut Wohnen und Umwelt GmbH, Büro für ökologische Energienutzung (BÖE), 1996 [viewed 20 March 2017]. Available from: http://www.iwu.de/fileadmin/user_upload/dateien/energie/neh_ph/BHKW_Jahresdauerlinien.pdf

[66] MOHR, J. and SCHARRE, T. Fernwärme Hamburg. In: *Fachveranstaltung Heizungsnetzwerk am 21.03.2013* [online]. Hamburg: Vattenfall Wärme Hamburg GmbH, 2013. [viewed 21 June 2015]. Available from: http://www.hamburg.de/contentblob/3902872/data/hn-vortrag-vattenfall-fernwaerme.pdf

[67] AGEB. *Bruttostromerzeugung in Deutschland nach Energieträgern im Jahr 2014 (in Terawattstunden)* [online]. Statista.com, 2014 [viewed 19 January 2015]. Available from: http://de.statista.com/statistik/daten/studie/197025/umfrage/bruttostromerzeugung-in-deutschland-nach-energietraegern/

[68] VATTENFALL EUROPE WÄRME AG. *Kurzbeschreibung zum Antrag auf Genehmigung gemäß § 4 Absatz 1 BImSchG für die Errichtung und den Betrieb des Gas- und Dampfturbinen- Heizkraftwerks Wedel am Standort Tinsdaler Weg 146, 22880 Wedel* [online]. Vattenfall Europe Wärme AG, 2012 [viewed 10 March 2013]. Available from: http://innovations-kraftwerk-wedel.de/ikwdaten/2013/01/20120806_HKW_Wedel_Kurzbeschreibung.pdf_22234222.pdf

[69] H2-NETZWERK-RUHR E. V. *Hydrogen and Fuel Cells in the Ruhr Region – Pillars for the Future of Energy and Industry* [online]. Herten: h2-netzwerk-ruhr e. V., 2013 [viewed 20 March 2017]. Available from: http://www.h2-netzwerk-ruhr.de/uploads/media/Brochure_englisch.pdf

[70] NOACK, C., BURGGRAF, F., SCHWAN, S., HOSSEINY, P.L., KOLB, S., BELZ, S., DONADEI, S., KALLO, J., FRIEDRICH, A., PREGGER, T., KIEN CAO, K., HEIDE, D. and NAEGLER, T. *Studie über die Planung einer Demonstrationsanlage zur Wasserstoff-Kraftstoffgewinnung durch Elektrolyse mit Zwischenspeicherung in Salzkavernen unter Druck* [online]. Stuttgart: Deutsches Zentrum für Luft- und Raumfahrt e. V., Ludwig-Bölkow-Systemtechnik GmbH, Fraunhofer ISE, KBB Underground Technologies GmbH, 2014 [viewed 20 March 2017]. Available from: http://elib.dlr.de/94979/1/2014_DLR_ISE_KBB_LBST_PlanDelyKaD.pdf

[71] VATTENFALL GMBH. *Spitzentechnologie im Einsatz - Kraftwerk Moorburg* [online]. Vattenfall.de, 2015 [viewed 22 July 2015]. Available from: http://corporate.vattenfall.de/uber-uns/geschaftsfelder/erzeugung/bauprojekte/Moorburg/Spitzentechnologie_im_Einsatz/

[72] KEHLHOFER, R., BACHMANN, R., NIELSEN, H. and WARNER, J. *Combined-Cycle Gas and Steam Turbine Power Plants*. 2nd Ed. Tulsa, Oklahoma: PennWell Publishing Company, 1999. ISBN 0878147365.

[73] KATHER, A., ROEDER, V., HASENBEIN, C., SCHMITZ, G., WELLNER, K., GOTTELT, F. and NIELSEN, L. *DYNCAP - Dynamische Untersuchung von Dampfkraftprozessen mit CO_2-Abtrenung zur Bereitstellung von Regelenergie. Abschlussbericht des Verbundvorhabens* [online]. Hamburg: Technische Universität Hamburg-Harburg, XRG Simulation GmbH, TLK Thermo GmbH, 2015 [viewed 20 March 2017]. Available from: https://www.tib.eu/suchen/id/TIBKAT:837510201/

[74] EUROPEAN PARLIAMENT. *Directive 2004/8/EC of the European Parliament and of the Council of 11 February 2004 on the promotion of cogeneration based on a useful heat demand in the internal energy market and amending Directive 92/42/EEC* [online]. Official Journal of the European Union, 2004 [viewed 20 March 2017]. Available from: http://eur-lex.europa.eu/legal-content/EN/TXT/PDF/?uri=CELEX:32004L0008&from=EN

[75] BEECKEN, J., RIDDER, M., SCHAPER, H., SCHÖTTKER, P., MICUS, W., ROGALLA, B.-U. and FEUERRIEGEL, S. *Bessere Ausnutzung von Fernwärmeanlagen, Teilprojekt Hannover-Hamburg: Analyse des Regelverhaltens von Fernwärmenetzen* [online]. Hamburg: Hamburgische Electricitäts-Werke AG, 2000 [viewed 20 March 2017]. Available from: https://www.tib.eu/de/suchen/download///?tx_tibsearch_search%5Bdocid%5D=TIBKAT%3A32977557X

[76] WIRZ, F. *Simulation zur Ermittlung einer Blockvertrimstrategie für das Heizkraftwerk Wedel - Kleine Studienarbeit.* Technische Universität Hamburg-Harburg, Arbeitsbereich Wärmekraftanlagen und Schiffsmaschinen, 2005.

[77] THIERFELDER, H.-G. and GROSSTERLINDE, F.-W. Das HEW-Heizkraftwerk Tiefstack - Erste Erfahrungen mit einer der modernsten und umweltfreundlichsten Anlagen. *VGB Kraftwerkstechnik.* 1995, Vol. 75, no. 4, p. 345–352

[78] HARTMANN, C. and SCHRADER, K. Aus Strom wird Wärme: Elektrodenheizkessel in Flensburg. *EuroHeat&Power.* 2014, Vol. 43, no. 4, p. 48–53

[79] KONSTANTIN, P. *Praxisbuch Energiewirtschaft.* Burgstetten: Springer-Verlag Berlin Heidelberg, 2013. ISBN 9783642372643.

[80] BÜRGER, V., HESSE, T., QUACK, D., PALZER, A., KÖHLER, B., HERKEL, S. and ENGELMANN, P. *Klimaneutraler Gebäudebestand 2050* [online]. Dessau-Roßlau: Umweltbundesamt, 2016 [viewed 20 March 2017]. Available from: https://www.umweltbundesamt.de/sites/default/files/medien/378/publikationen/climate_change_06_2016_klimaneutraler_gebaeudebestand_2050.pdf

[81] RODEWALD, A. *Kohlendioxidemissionsszenarien unter Berücksichtigung zukünftiger Stromerzeugungstechnologien.* Düsseldorf: VDI-Verlag, 2008. ISBN 9783183578061.

[82] KRIEG, D. *Konzept und Kosten eines Pipelinesystems zur Versorgung des deutschen Straßenverkehrs mit Wasserstoff* [online]. Jülich: Forschungszentrum Jülich GmbH, 2010 [viewed 20 March 2017]. Available from: http://juser.fz-juelich.de/record/136392/files/Energie%26Umwelt_144.pdf

[83] SCHMITT, F., CASPAR, J., HOLLER, S. and KLÖPSCH, M. *Wärmetransport im Wettbewerb zu dislozierter Wärmeerzeugung* [online]. Mannheim: MVV Energie AG, 2013 [viewed 20 March 2017]. Available from: http://www.eneff-stadt.info/fileadmin/media/Projektbilder/Neue_Technologien/Kostenguenstiger_Fernwaermetransport_Mvv/Abschlussbericht__Teilprojekt_1_Waermetransport_im_Wettbewerb_2014_10_07.pdf

[84] FINANZEN.NET GMBH. *Rohstoffe* [online]. Finanzen.net, 2016 [viewed 1 June 2016]. Available from: http://www.finanzen.net/rohstoffe

[85] GERMAN BUNDESTAG. *Kraft-Wärme-Kopplungsgesetz vom 21. Dezember 2015 (BGBl. I S. 2498)* [online]. Bundesministerium der Justiz und für Verbraucherschutz, 2017 [viewed 20 March 2017]. Available from: https://www.gesetze-im-internet.de/bundesrecht/kwkg_2016/gesamt.pdf

[86] BATEN, T., BUTTERMANN, H. and NIEDER, T. Gesamtbilanz der Kraft-Wärme-Kopplung 2003 bis 2012. *Energiewirtschaftliche Tagesfragen* [online]. 2014, Vol. 64, no. 5, p. 37–44 [viewed 20 March 2017]. Available from: http://eefa.de/Baten_et_al_ET_5_2014.pdf

[87] TIETJE, H. and TEUNIS, S.-L. *Energiebilanz und CO_2 - Bilanzen für Hamburg 2012* [online]. Hamburg: Freie und Hansestadt Hamburg - Behörde für Stadtentwicklung und Umwelt, 2012 [viewed 20 March 2017]. Available from: http://www.statistik-nord.de/fileadmin/Dokumente/Sonderveröffentlichungen/Energiebilanz_Hamburg/EB_CO2_HH_2012.pdf

Lebenslauf

Persönliche Daten

Name:	Peniche Garcia
Vorname:	Ricardo
Geburtstag:	08. Juni 1984
Geburtsort:	Merida, Mexiko

Schulischer Werdegang

09/1990 - 06/2002	Grundschule und Gymnasium Colegio Montejo / Centro Universitario Montejo Merida, Mexiko

Studium

08/2003 – 12/2008	Tecnologico de Monterrey (ITESM) Abschluss: Ingeniero Mecanico Electricista
08/2007 – 03/2008	Technische Universität Hamburg-Harburg Austauschstudent und Praktikant
10/2009 – 06/2012	Technische Universität Berlin Abschluss: M.Sc. Global Production Engineering

Beruflicher Werdegang

01/2009 – 08/2009	Design- und Feldingenieur Diram SA de CV
06/2012 – 11/2012	Ingenieur für Fertigungstechnik Solarlite GmbH
05/2013 – 04/2017	Wissenschaftlicher Mitarbeiter Institut für Energietechnik Technische Universität Hamburg-Harburg